CAMBRIDGE LIBRARY COLLECTION

Books of enduring scholarly value

Earth Sciences

In the nineteenth century, geology emerged as a distinct academic discipline. It pointed the way towards the theory of evolution, as scientists including Gideon Mantell, Adam Sedgwick, Charles Lyell and Roderick Murchison began to use the evidence of minerals, rock formations and fossils to demonstrate that the earth was older by millions of years than the conventional, Bible-based wisdom had supposed. They argued convincingly that the climate, flora and fauna of the distant past could be deduced from geological evidence. Volcanic activity, the formation of mountains, and the action of glaciers and rivers, tides and ocean currents also became better understood. This series includes landmark publications by pioneers of the modern earth sciences, who advanced the scientific understanding of our planet and the processes by which it is constantly re-shaped.

Geological Observations on the Volcanic Islands, Visited During the Voyage of H.M.S. Beagle

Charles Darwin (1809–1882) published *Observations on the Volcanic Islands* in 1844. It is one of three major geological works resulting from the voyage of the Beagle, and contains detailed geological descriptions of locations visited by Darwin including the Cape Verde archipelago, Mauritius, Ascension Island, St Helena, the Galápagos, and parts of Australia, New Zealand and South Africa. Chapter 6 discusses the types of lava found on different oceanic islands. There is an appendix of short contributions by two other scholars: descriptions of fossil shells from Cape Verde, St Helena and Tasmania by G. B. Sowerby and of fossil corals from Tasmania by W. Lonsdale. The book is illustrated with woodcuts, maps and sketches of specimens. It provides valuable insights into one of the most important scientific voyages ever made, and the development of Darwin's ideas on geology.

Cambridge University Press has long been a pioneer in the reissuing of out-of-print titles from its own backlist, producing digital reprints of books that are still sought after by scholars and students but could not be reprinted economically using traditional technology. The Cambridge Library Collection extends this activity to a wider range of books which are still of importance to researchers and professionals, either for the source material they contain, or as landmarks in the history of their academic discipline.

Drawing from the world-renowned collections in the Cambridge University Library, and guided by the advice of experts in each subject area, Cambridge University Press is using state-of-the-art scanning machines in its own Printing House to capture the content of each book selected for inclusion. The files are processed to give a consistently clear, crisp image, and the books finished to the high quality standard for which the Press is recognised around the world. The latest print-on-demand technology ensures that the books will remain available indefinitely, and that orders for single or multiple copies can quickly be supplied.

The Cambridge Library Collection will bring back to life books of enduring scholarly value (including out-of-copyright works originally issued by other publishers) across a wide range of disciplines in the humanities and social sciences and in science and technology.

Geological Observations on the Volcanic Islands, Visited During the Voyage of H.M.S. Beagle

*Together with Some Brief Notices
on the Geology of Australia and
the Cape of Good Hope*

CHARLES DARWIN

CAMBRIDGE
UNIVERSITY PRESS

CAMBRIDGE UNIVERSITY PRESS

Cambridge, New York, Melbourne, Madrid, Cape Town,
Singapore, São Paolo, Delhi, Tokyo, Mexico City

Published in the United States of America by Cambridge University Press, New York

www.cambridge.org
Information on this title: www.cambridge.org/9781108072335

This edition first published 1844
This digitally printed version 2011

ISBN 978-1-108-07233-5 Paperback

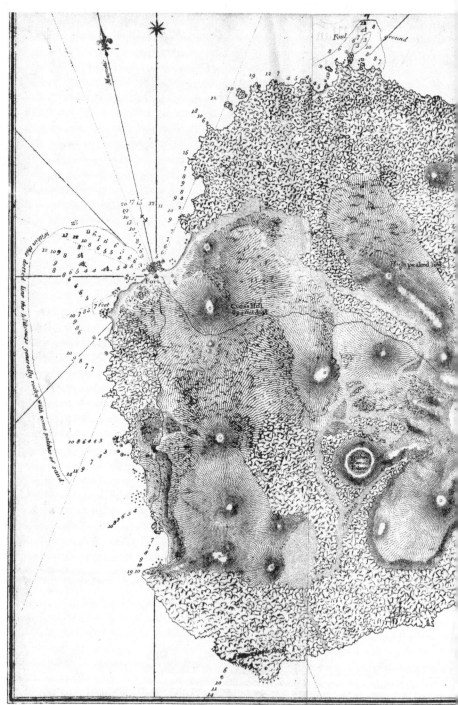

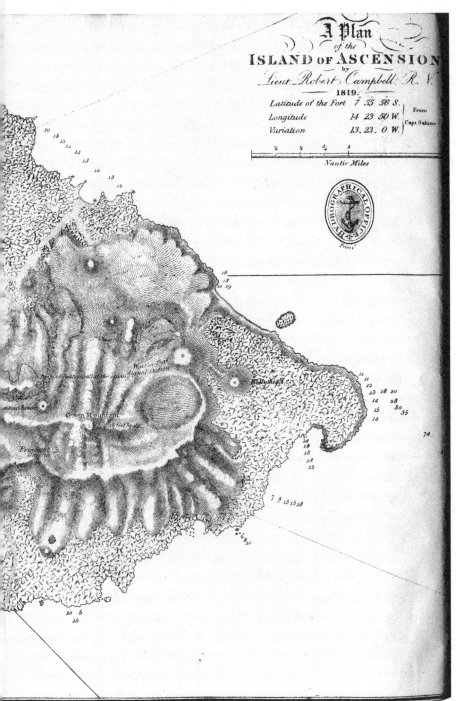

A Plan
of the
ISLAND OF ASCENSION
by
Lieut. Robert Campbell. R.N.
1819.

		From
Latitude of the Fort	7 55 56 S.	
Longitude	14 23 50 W.	Capt. Sabine
Variation	13. 23. 0 W.	

Nautic Miles

HYDROGRAPHICAL OFFICE

Price 1.

Weather Post
Signal Station

Holly Head

Green Mountain

the Hydrographical Office of the Admiralty. 14.th Sep.r 1825.

J. Walker

GEOLOGICAL OBSERVATIONS

ON THE

VOLCANIC ISLANDS,

VISITED DURING THE VOYAGE OF H. M. S. BEAGLE,

TOGETHER WITH

SOME BRIEF NOTICES ON THE GEOLOGY OF AUSTRALIA AND
THE CAPE OF GOOD HOPE.

BEING THE SECOND PART OF
THE GEOLOGY OF THE VOYAGE OF THE BEAGLE,
UNDER THE COMMAND OF CAPT. FITZROY, R.N.
DURING THE YEARS 1832 TO 1836.

BY

CHARLES DARWIN, M.A., F.R.S.,

VICE-PRESIDENT OF THE GEOLOGICAL SOCIETY, AND NATURALIST TO THE EXPEDITION.

Published with the Approval of the Lords Commissioners of
Her Majesty's Treasury.

LONDON:
SMITH, ELDER AND CO., 65, CORNHILL.

1844.

CONTENTS.

CHAPTER IV.

ST. HELENA.

CHAPTER V.

GALAPAGOS ARCHIPELAGO.

CHAPTER VI.

TRACHYTE AND BASALT.—DISTRIBUTION OF VOLCANIC ISLES.

CHAPTER VII.

APPENDIX.

DESCRIPTION OF FOSSIL SHELLS, BY G. B. SOWERBY, ESQ., F.L.S.

DESCRIPTION OF FOSSIL CORALS FROM THE PALÆOZOIC FORMATION OF VAN DIEMEN'S LAND, BY W. LONSDALE, ESQ., F.G.S.

APPENDIX

CHAPTER I.

ST. JAGO, IN THE CAPE DE VERDE ARCHIPELAGO.

Rocks of the lowest series.—A calcareous sedimentary deposit, with recent shells, altered by the contact of superincumbent lava, its horizontality and extent—Subsequent volcanic eruptions, associated with calcareous matter in an earthy and fibrous form, and often enclosed within the separate cells of the scoriæ—Ancient and obliterated orifices of eruption of small size—Difficulty of tracing over a bare plain recent streams of lava—Inland hills of more ancient volcanic rock—Decomposed olivine in large masses—Feldspathic rocks beneath the upper crystalline basaltic strata—Uniform structure and form of the more ancient volcanic hills—Form of the valleys near the coast—Conglomerate now forming on the sea beach.

THE island of St. Jago extends in a N.N.W. and S.S.E. direction, thirty miles in length by about twelve in breadth. My observations, made during two visits, were confined to the southern portion within the distance of a few leagues from Porto Praya. The country viewed from the sea, pre-

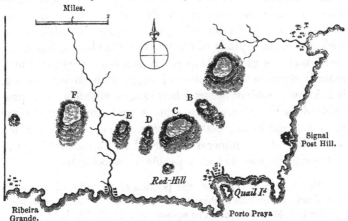

PART OF ST. JAGO, ONE OF THE CAPE DE VERDE ISLANDS.

B

sents a varied outline : smooth conical hills of a reddish
colour (like Red Hill in the accompanying wood-cut),* and
othersless regular, flat-topped, and of a blackish colour (like
A, B, C,) rise from successive, step-formed plains of lava.
At a distance, a chain of mountains, many thousand feet in
height, traverses the interior of the island. There is no
active volcano in St. Jago, and only one in the group,
namely at Fogo. The island since being inhabited, has not
suffered from destructive earthquakes.

The lowest rocks exposed on the coast near Porto Praya,
are highly crystalline and compact; they appear to be of
ancient, submarine, volcanic origin; they are unconformably
covered by a thin, irregular, calcareous deposit, abounding
with shells of a late tertiary period; and this again is capped
by a wide sheet of basaltic lava, which has flowed in successive streams from the interior of the island, between the
square-topped hills marked A, B, C, &c. Still more recent
streams of lava have been erupted from the scattered cones,
such as Red and Signal Post Hills. The upper strata of the
square-topped hills are intimately related in mineralogical
composition, and in other respects, with the lowest series of
the coast-rocks, with which they seem to be continuous.

Mineralogical description of the rocks of the lowest series.—
These rocks possess an extremely varying character; they
consist of black, brown and gray, compact, basaltic bases,
with numerous crystals of augite, hornblende, olivine, mica,
and sometimes glassy feldspar. A common variety is almost
entirely composed of crystals of augite with olivine. Mica,
it is known, seldom occurs where augite abounds; nor probably does the present case offer a real exception, for the
mica (at least in my best characterized specimen, in which
one nodule of this mineral is nearly half an inch in length,)
is as perfectly rounded as a pebble in a conglomerate, and

* The outline of the coast, the position of the villages, streamlets,
and of most of the hills in this wood-cut, are copied from the chart made
on board H. M. S. Leven. The square topped hills (A, B, C, &c.) are
put in merely by eye, to illustrate my description.

evidently has not been crystallized in the base, in which it is now inclosed, but has proceeded from the fusion of some pre-existing rock. These compact lavas alternate with tuffs, amygdaloids and wacke, and in some places with coarse conglomerate. Some of the argillaceous wackes are of a dark green colour, others, pale yellowish-green, and others nearly white; I was surprised to find that some of the latter varieties, even where whitest, fused into a jet black enamel, whilst some of the green varieties afforded only a pale gray bead. Numerous dikes, consisting chiefly of highly compact augitic rocks, and of gray amygdaloidal varieties, intersect the strata, which have in several places been dislocated with considerable violence, and thrown into highly-inclined positions. One line of disturbance, crosses the northern end of Quail Island, (an islet in the bay of Porto Praya) and can be followed to the mainland. These disturbances took place before the deposition of the recent sedimentary bed; and the surface, also, had previously been denuded to a great extent, as is shown by many truncated dikes.

Description of the calcareous deposit overlying the fore-going volcanic rocks. — This stratum is very conspicuous from its white colour, and from the extreme regularity with which it ranges in a horizontal line for some miles along the coast. Its average height above the sea, measured from the upper line of junction with the superincumbent basaltic lava, is about sixty feet; and its thickness, although varying much from the inequalities of the underlying formation, may be estimated at about twenty feet. It consists of quite white calcareous matter, partly composed of organic debris, and partly of a substance which may be aptly compared in appearance with mortar. Fragments of rock and pebbles are scattered throughout this bed, often forming, especially in the lower part, a conglomerate. Many of the fragments of rock are whitewashed with a thin coating of calcareous matter. At Quail Island, the calcareous deposit is replaced in its lowest part by a soft, brown, earthy tuff, full of Turritellæ;

this is covered by a bed of pebbles, passing into sandstone, and mixed with fragments of echini, claws of crabs, and shells; the oyster shells still adhering to the rock on which they grew. Numerous white balls appearing like pisolitic concretions, from the size of a walnut to that of an apple, are embedded in this deposit; they usually have a small pebble in their centres. Although so like concretions, a close examination convinced me that they were Nulliporæ, retaining their proper forms, but with their surfaces slightly abraded: these bodies (plants as they are now generally considered to be), exhibit under a microscope of ordinary power, no traces of organization in their internal structure. Mr. George R. Sowerby has been so good as to examine the shells which I collected: there are fourteen species in a sufficiently perfect condition for their characters to be made out with some degree of certainty, and four which can be referred only to their genera. Of the fourteen shells, of which a list is given in the Appendix, eleven are recent species; one, though undescribed, is perhaps identical with a species, which I found living in the harbour of Porto Praya; the two remaining species are unknown, and have been described by Mr. Sowerby. Until the shells of this Archipelago and of the neighbouring coasts are better known, it would be rash to assert that even these two latter shells are extinct. The number of species which certainly belong to existing kinds, although few in number, are sufficient to show that the deposit belongs to a late tertiary period. From its mineralogical character, from the number and size of the embedded fragments, and from the abundance of Patellæ, and other littoral shells, it is evident that the whole was accumualted in a shallow sea, near an ancient coast-line.

Effects produced by the flowing of the superincumbent basaltic lava over the calcareous deposit.—These effects are very curious. The calcareous matter is altered to the depth of about a foot beneath the line of junction; and a most perfect gradation can be traced, from loosely aggregated, small, particles of shells, corrallines, and Nulliporæ, into a rock, in which not a trace of mechanical origin can be discovered, even with a microscope. Where the metamorphic change has been greatest, two varieties occur. The first is a hard, compact, white, fine grained rock, striped with a few parallel lines of black volcanic particles, and resembling a sandstone, but which, upon close examination, is seen to be crystallized throughout, with the cleavages so perfect that they can be readily measured by the reflecting goniometer. In specimens, where the change has been less complete, when moistened and examined under a strong lens, the most interesting gradation can be traced, some of the rounded particles retaining their proper forms, and others insensibly melting into the granulo-crystalline paste. The weathered surface of this stone, as is so frequently the case with ordinary limestones, assumes a brick-red colour.

The second metamorphosed variety is likewise a hard rock, but without any crystalline structure. It consists of a white, opaque, compact, calcareous stone, thickly mottled with rounded, though irregular, spots of a soft, earthy, ochraceous substance. This earthy matter is of a pale yellowish-brown colour, and appears to be a mixture of carbonate of lime with iron; it effervesces with acids, is infusible, but blackens under the blow-pipe, and becomes magnetic. The rounded form of the minute patches of earthy substance, and the steps in the progress of their perfect formation, which can be followed in a suit of specimens, clearly show that they are due either to some power of aggregation in the earthy particles amongst themselves, or more probably to a strong attraction between the atoms of the carbonate of lime, and consequently to the segregation of the earthy extraneous matter. I was much interested by this fact, because I have often seen quartz rocks (for instance, in the Falkland Islands, and in the lower Silurian strata of the Stiper-stones in Shropshire), mottled in a precisely analogous manner, with little spots of a white, earthy substance (earthy feldspar?); and these rocks, there was good reason to suppose, had undergone the action of heat,—a view which thus receives confirmation.

This spotted structure may possibly afford some indication
in distinguishing those formations of quartz, which owe their
present structure to igneous action, from those produced by
the agency of water alone ; a source of doubt, which I should
think from my own experience, that most geologists, when
examining arenaceo-quartzose districts, must have experienced.

The lowest and most scoriaceous part of the lava, in rolling
over the sedimentary deposit at the bottom of the sea, has
caught up large quantities of calcareous matter, which now
forms a snow-white, highly crystalline, basis to a breccia, including small pieces of black, glossy scoriæ.   A little above
this, where the lime is less abundant, and the lava more compact, numerous little balls, composed of spicula of calcareous
spar, radiating from common centres, occupy the interstices.
In one part of Quail Island, the lime has thus been crystallized by the heat of the superincumbent lava, where it is
only thirteen feet in thickness ; nor had the lava been
originally thicker, and since reduced by degradation, as
could be told from the degree of cellularity of its surface.
I have already observed that the sea must have been shallow
in which the calcareous deposit was accumulated.   In this
case, therefore, the carbonic acid gas has been retained
under a pressure, insignificant compared with that (a column
of water, 1708 feet in height) originally supposed by Sir
James Hall to be requisite for this end : but since his experiments, it has been discovered that pressure has less to do
with the retention of carbonic acid gas, than the nature
of the circumjacent atmosphere ; and hence, as is stated to
be the case by Mr. Faraday,* masses of limestone are sometimes fused and crystallized even in common lime-kilns.
Carbonate of lime can be heated to almost any degree,

---

* I am much indebted to Mr. E. W. Brayley in having given me the
following references to papers on this subject : Faraday, in the Edinburgh New Philosophical Journal, vol. xv. p. 398; Gay Lussac, in
Annales de Chem. et Phys. tom. lxiii. p. 219, translated in the London
and Edinburgh Philosophical Magazine, vol. x. p. 496.

according to Faraday, in an atmosphere of carbonic acid gas, without being decomposed; and Gay Lussac found that fragments of limestone, placed in a tube and heated to a degree, not sufficient by itself to cause their decomposition, yet immediately evolved their carbonic acid, when a stream of common air or steam was passed over them : Gay Lussac attributes this to the mechanical displacement of the nascent carbonic acid gas. The calcareous matter beneath the lava, and especially that forming the crystalline spicula between the interstices of the scoriæ, although heated in an atmosphere probably composed chiefly of steam, could not have been subjected to the effects of a passing stream; and hence it is, perhaps, that they have retained their carbonic acid, under a small amount of pressure.

The fragments of scoriæ, embedded in the crystalline calcareous basis, are of a jet black colour, with a glossy fracture like pitchstone. Their surfaces, however, are coated with a layer of a reddish-orange, translucent substance, which can easily be scratched with a knife; hence they appear as if overlaid by a thin layer of rosin. Some of the smaller fragments are partially changed throughout into this substance: a change which appears quite different from ordinary decomposition. At the Galapagos Archipelago (as will be described in a future chapter,) great beds are formed of volcanic ashes and particles of scoriæ, which have undergone a closely similar change.

The extent and horizontality of the calcareous stratum.—The upper line of surface of the calcareous stratum, which is so conspicuous from being quite white and so nearly horizontal, ranges for miles along the coast, at the height of about sixty feet above the sea. The sheet of basalt, by which it is capped, is on an average eighty feet in thickness. Westward of Porto Praya beyond Red Hill, the white stratum with the superincumbent basalt is covered up by more recent streams. Northward of Signal Post Hill, I could follow it with my eye, trending away for several miles along the sea cliffs. The distance thus observed is about seven miles; but I cannot doubt from its regularity, that it extends much further. In some ravines at right angles to the coast, it is seen gently dipping towards the sea, probably with the same inclination as when deposited round the ancient shores of the island. I found only one inland section, namely, at the base of the hill marked A, where, at the height of some hundred feet, this bed was exposed; it here rested on the usual compact augitic rock associated with wacke, and was covered by the wide-spread sheet of modern basaltic lava. Some exceptions occur to the horizontality of the white stratum: at Quail Island, its upper surface is only forty feet above the level of the sea; here also the capping of lava is only between twelve and fifteen feet in thickness; on the other hand, at the N.E. side of Porto Praya harbour, the calcareous stratum, as well as the rock on which it rests, attain a height above the average level: the inequality of level in these two cases is not, as I believe, owing to unequal elevation, but to original irregularities at the bottom of the sea. Of this fact, at Quail Island, there was clear evidence in the calcareous deposit being in one part of much greater than the average thickness, and in another part being entirely absent; in this latter case, the modern basaltic lavas rested directly on those of more ancient origin.

Under Signal Post Hill, the white stratum dips into the sea in a remarkable manner. This hill is conical, 450 feet in height, and retains some traces of having had a crateriform structure; it is composed chiefly of matter erupted posteriorly to the elevation of the great basaltic plain, but partly of lava of apparently submarine origin and of considerable antiquity. The surrounding plain, as well as the eastern flank of this hill, have been worn into steep precipices, over-hanging the sea. In these precipices, the white calcareous stratum may be seen, at the height of about 70 feet above the beach, running for some miles both northward and southward of the hill, in a line appearing to be perfectly horizontal: but for a space of a quarter of a mile directly under the hill, it dips into the sea and disappears. On the south side the dip is gradual, on the north side it is more abrupt, as is shown in the woodcut. As neither the cal-

No. 2.

SIGNAL POST HILL.

A—Ancient volcanic rocks.
B—Calcareous stratum.
C—Upper basaltic lava.

careous stratum, nor the superincumbent basaltic lava (as far as the latter can be distinguished from the more modern ejections), appear to thicken as they dip, I infer that these strata were not originally accumulated in a trough, the centre of which afterwards became a point of eruption; but that they have subsequently been disturbed and bent. We may suppose either that Signal Post Hill subsided after its elevation with the surrounding country, or that it never was uplifted to the same height with it. This latter seems to me the most probable alternative, for during the slow and equable elevation of this portion of the island, the subterranean motive power, from expending part of its force, in repeatedly erupting volcanic matter from beneath this point, would, it is likely, have less force to uplift it. Something of the same kind seems to have occurred near Red Hill, for when tracing upwards the naked streams of lava from near Porto Praya towards the interior of the island, I was strongly induced to suspect, that since the lava had flowed, the slope of the land had been slightly modified, either by a small subsidence near Red Hill, or by that portion of the plain having been uplifted to a less height during the elevation of the whole area.

The basaltic lava, superincumbent on the calcareous deposit. — This lava is of a pale gray colour, fusing into a black enamel; its fracture is rather earthy and concretionary; it contains olivine in small grains. The central parts of the mass are compact, or at most crenulated with a few minute cavities, and are often columnar. At Quail Island this structure was assumed in a striking manner; the lava in one part being divided into horizontal laminæ, which became in another part split by vertical fissures into five-sided plates; and these again, being piled on each other, insensibly became soldered together, forming fine symmetrical columns. The lower surface of the lava is visicular, but sometimes only to the thickness of a few inches; the upper surface, which is likewise vesicular, is divided into balls, frequently as much as three feet in diameter, made up of concentric layers. The mass is composed of more than one stream; its total thickness being, on an average, about eighty feet: the lower portion has certainly flowed beneath the sea, and probably likewise the upper portion. The chief part of this lava has flowed from the central districts, between the hills marked A, B, C, &c. in the woodcut-map. The surface of the country, near the coast, is level and barren; towards the interior, the land rises by successive terraces, of which four, when viewed from a distance, could be distinctly counted.

Volcanic eruptions subsequent to the elevation of the coast-land; the ejected matter associated with earthy lime.—These recent lavas have proceeded from those scattered, conical, reddish-coloured hills, which rise abruptly from the plain-country near the coast. I ascended some of them, but will describe only one, namely, *Red Hill*, which may serve as a type of its class, and is remarkable in some especial respects. Its height is about 600 feet; it is composed of bright red, highly scoriaceous rock of a basaltic nature; on one side of its summit there is a hollow, probably the last remnant of a crater. Several of the other hills of this class, judging from their external forms, are surmounted by much more perfect craters. When sailing along the coast, it was evident that a considerable body of lava had flowed from Red Hill, over a line of cliff about 120 feet in height, into the sea: this line of cliff is continuous with that forming the coast, and bounding the plain on both sides of this hill; these streams, therefore, were erupted, after the formation of the coast-cliffs, from Red Hill, when it must have stood, as it now does, above the

level of the sea. This conclusion accords with the highly
scoriaceous condition of all the rock on it, appearing to be
of subaërial formation; and this is important, as there are
some beds of calcareous matter near its summit, which might,
at a hasty glance, have been mistaken for a submarine deposit. These beds consist of white, earthy, carbonate of
lime, extremely friable, so as to be crushed with the least
pressure; the most compact specimens not resisting the
strength of the fingers. Some of the masses are as white as
quick-lime, and appear absolutely pure; but on examining
them with a lens, minute particles of scoriæ can always be
seen, and I could find none which, when dissolved in acids,
did not leave a residue of this nature. It is, moreover, difficult to find a particle of the lime which does not change
colour under the blowpipe, most of them even becoming
glazed. The scoriaceous fragments and the calcareous
matter are associated in the most irregular manner, sometimes in obscure beds, but more generally as a confused
breccia, the lime in some parts and the scoriæ in others
being most abundant. Sir H. De la Beche has been so kind
as to have some of the purest specimens analyzed, with a
view to discover, considering their volcanic origin, whether
they contained much magnesia; but only a small portion
was found, such as is present in most limestones.

Fragments of the scoriæ embedded in the calcareous mass,
when broken, exhibit many of their cells lined and partly
filled with a white, delicate, excessively fragile, moss-like, or
rather conferva-like, reticulation of carbonate of lime. These
fibres, examined under a lens of one-tenth of an inch focal
distance, appear cylindrical; they are rather above the $\frac{1}{1000}$
of an inch in diameter; they are either simply branched, or
more commonly united into an irregular mass of net-work,
with the meshes of very unequal sizes and of unequal numbers of sides. Some of the fibres are thickly covered with
extremely minute spicula, occasionally aggregated into little
tufts; and hence they have a hairy appearance. These
spicula are of the same diameter throughout their length;

they are easily detached, so that the object-glass of the microscope soon becomes scattered over with them. Within the cells of many fragments of the scoriæ, the lime exhibits this fibrous structure, but generally in a less perfect degree. These cells do not appear to be connected with one another. There can be no doubt, as will presently be shown, that the lime was erupted, mingled with the lava in its fluid state; and therefore I have thought it worth while to describe minutely this curious fibrous structure, of which I know nothing analogous. From the earthy condition of the fibres, this structure does not appear to be related to crystallization.

Other fragments of the scoriaceous rock from this hill, when broken, are often seen marked with short and irregular white streaks, which are owing to a row of separate cells being partly, or quite, filled with white calcareous powder. This structure immediately reminded me of the appearance in badly kneaded dough, of balls and drawn-out streaks of flour, which have remained unmixed with the paste; and I cannot doubt that small masses of the lime, in the same manner remaining unmixed with the fluid lava, have been drawn out, when the whole was in motion. I carefully examined, by trituration and solution in acids, pieces of the scoriæ, taken from within half-an-inch of those cells which were filled with the calcareous powder, and they did not contain an atom of free lime. It is obvious that the lava and lime have on a large scale been very imperfectly mingled; and where small portions of the lime have been entangled within a piece of the viscid lava, the cause of their now occupying, in the form of a powder or of a fibrous reticulation, the vesicular cavities, is, I think, evidently due to the confined gases having most readily expanded at the points, where the incoherent lime rendered the lava less adhesive.

A mile eastward of the town of Praya, there is a steep-sided gorge, about 150 yards in width, cutting through the basaltic plain and underlying beds, but since filled up by a stream of more modern lava. This lava is dark gray, and in most parts compact and rudely columnar; but at a little distance from the coast, it includes in an irregular manner, a brecciated mass of red scoriæ mingled with a considerable quantity of white, friable, and in some parts, nearly pure earthy lime, like that on the summit of Red Hill. This lava, with its entangled lime, has certainly flowed in the form of a regular stream; and, judging from the shape of the gorge, towards which the drainage of the country (feeble though it now be) still is directed, and from the appearance of the bed of loose water-worn blocks with their interstices unfilled, like those in the bed of a torrent, on which the lava rests, we may conclude that the stream was of subaërial origin. I was unable to trace it to its source, but, from its direction, it seemed to have come from Signal Post Hill, distant one mile and a quarter, which, like Red Hill, has been a point of eruption subsequently to the elevation of the great basaltic plain. It accords with this view, that I found on Signal Post Hill, a mass of earthy, calcareous matter of the same nature, mingled with scoriæ. I may here observe that part of the calcareous matter forming the horizontal sedimentary bed, especially the finer matter with which the embedded fragments of rock are whitewashed, has probably been derived from similar volcanic eruptions, as well as from triturated organic remains: the underlying, ancient, crystalline rocks, also, are associated with much carbonate of lime, filling amygdaloidal cavities, and forming irregular masses, the nature of which latter I was unable to understand.

Considering the abundance of earthy lime near the summit of Red Hill, a volcanic cone 600 feet in height, of subaërial growth,—considering the intimate manner in which minute particles and large masses of scoriæ are embedded in the masses of nearly pure lime, and on the other hand, the manner in which small kernels and streaks of the calcareous powder are included in solid pieces of the scoriæ,—considering, also, the similar occurrence of lime and scoriæ within a stream of lava, also supposed, with good reason, to have

been of modern subaërial origin, and to have flowed from a
hill, where earthy lime also occurs: I think, considering
these facts, there can be no doubt that the lime has been
erupted, mingled with the molten lava. I am not aware
that any similar case has been described: it appears to me an
interesting one, inasmuch as most geologists must have speculated on the probable effects of a volcanic focus, bursting
through deep-seated beds of different mineralogical composition. The great abundance of free silex in the trachytes of
some countries (as described by Beudant in Hungary, and
by P. Scrope in the Panza Islands), perhaps solves the enquiry with respect to deep-seated beds of quartz; and we
probably here see it answered, where the volcanic action has
invaded subjacent masses of limestone. One is naturally led
to conjecture, in what state the now earthy carbonate of lime
existed, when ejected with the intensely heated lava: from
the extreme cellularity of the scoriæ on Red Hill, the pressure cannot have been great, and as most volcanic eruptions
are accompanied by the emission of large quantities of steam
and other gases, we here have the most favourable conditions,
according to the views at present entertained by chemists, for
the expulsion of the carbonic acid.* Has the slow re-ab-
sorption of this gas, it may be asked, given to the lime in the
cells of the lava, that peculiar fibrous structure, like that of
an efflorescing salt? Finally, I may remark on the great
contrast in appearance between this earthy lime, which must
have been heated in a free atmosphere of steam and other
gases, with the white, crystalline, calcareous spar, produced

* Whilst deep beneath the surface, the carbonate of lime was, I
presume, in a fluid state. Hutton, it is known, thought that all amygdaloids were produced by drops of molten limestone floating in the
trap, like oil in water: this no doubt is erroneous, but if the matter
forming the summit of Red Hill had been cooled under the pressure of
a moderately deep sea, or within the walls of a dike, we should, in all
probability, have had a trap rock associated with large masses of compact, crystalline, calcareous spar, which, according to the views entertained by many geologists, would have been wrongly attributed to subsequent infiltration.

by a single thin sheet of lava (as at Quail Island) rolling over similar earthy lime and the debris of organic remains, at the bottom of a shallow sea.

Signal Post Hill.—This hill has already been several times mentioned, especially with reference to the remarkable manner in which the white calcareous stratum, in other parts so horizontal, (Wood-cut, No. 2.) dips under it into the sea. It has a broad summit, with obscure traces of a crateriform structure, and is composed of basaltic rocks,* some compact, others highly cellular, with inclined beds of loose scoriæ, of which some are associated with earthy lime. Like Red Hill it has been the source of eruptions, subsequently to the elevation of the surrounding basaltic plain; but unlike that hill, it has undergone considerable denudation, and has been the seat of volcanic action at a remote period, when beneath the sea. I judge of this latter circumstance from finding on its inland flank, the last remnants of three small points of eruption. These points are composed of glossy scoriæ, cemented by crystalline calcareous spar, exactly like the great submarine calcareous deposit, where the heated lava has rolled over it: their demolished state can, I think, be explained only by the denuding action of the waves of the sea. I was guided to the first orifice by observing a sheet of lava, about 200 yards square, with steepish sides, superimposed on the basaltic plain, with no adjoining hillock, whence it could have been erupted; and the only trace of a crater which I was able to discover, consisted of some inclined beds of scoriæ at one of its corners. At the distance of fifty yards from a second level-topped patch of

* Of these, one common variety is remarkable for being full of small fragments of a dark jasper-red earthy mineral, which, when examined carefully, shows an indistinct cleavage; the little fragments are elongated in form, are soft, are magnetic before and after being heated, and fuse with difficulty into a dull enamel. This mineral is evidently closely related to the oxides of iron, but I cannot ascertain what it exactly is. The rock containing this mineral, is crenulated with small angular cavities, which are lined and filled with yellowish crystals of carbonate of lime.

lava, but of much smaller size, I found an irregular circular group of masses of cemented, scoriaceous breccia, about six feet in height, which doubtless had once formed the point of eruption. The third orifice is now marked only by an irregular circle of cemented scoriæ, about four yards in diameter, and rising in its highest point scarcely three feet above the level of the plain, the surface of which, close all round, exhibits its usual appearance: here we have a horizontal basal section of a volcanic spiracle, which, together with all its ejected matter, has been almost totally obliterated.

The stream of lava, which fills the narrow gorge* eastward of the town of Praya, judging from its course, seems, as before remarked, to have come from Signal Post Hill, and to have flowed over the plain, after its elevation : the same observation applies to a stream, (possibly part of the same one,) capping the sea cliffs, a little eastward of the gorge. When I endeavoured to follow these streams over the stony level plain, which is almost destitute of soil and vegetation, I was much surprised to find, that although composed of hard basaltic matter, and not having been exposed to marine denudation, all distinct traces of them soon became utterly lost. But I have since observed at the Galapagos Archipelago, that it is often impossible to follow even great deluges of quite recent lava across older streams, except by the size of the bushes growing on them, or by the comparative states of glossiness of their surfaces, — characters which a short lapse of time would be sufficient quite to obscure. I may remark, that in a level country, with a dry climate, and with the wind blowing always in one direction, (as at the Cape de Verde Archipelago,) the effects of atmospheric degradation is probably much greater than would at first be expected; for soil in this case accumulates only in

---

* The sides of this gorge, where the upper basaltic stratum is intersected, are almost perpendicular. The lava, which has since filled it up, is attached to these sides, almost as firmly as a dike is to its walls. In most cases, where a stream of lava has flowed down a valley, it is bounded on each side by loose scoriaceous masses.

a few protected hollows, and being blown in one direction, it is always travelling towards the sea in the form of the finest dust, leaving the surface of the rocks bare, and exposed to the full effects of renewed meteoric action.

Inland hills of more ancient volcanic rocks.—These hills are laid down by eye, and marked as A, B, C, &c., in the wood-cut-map. They are related in mineralogical composition, and are probably directly continuous with the lowest rocks exposed on the coast. These hills, viewed from a distance, appear as if they had once formed part of an irregular table-land, and from their corresponding structure and composition this probably has been the case. They have flat, slightly inclined summits, and are, on an average, about 600 feet in height; they present their steepest slope towards the interior of the island, from which point they radiate outwards, and are separated from each other by broad and deep valleys, through which the great streams of lava, forming the coast-plains, have descended. Their inner and steeper escarpements are ranged in an irregular curve, which rudely follows the line of the shore, two or three miles inland from it. I ascended a few of these hills, and from others, which I was able to examine with a telescope, I obtained specimens, through the kindness of Mr. Kent, the assistant-surgeon of the *Beagle,* although by these means I am acquainted with only a part of the range, five or six miles in length; yet I scarcely hesitate, from their uniform structure, to affirm, that they are parts of one great formation, stretching round much of the circumference of the island.

The upper and lower strata of these hills differ greatly in composition. The upper are basaltic, generally compact, but sometimes scoriaceous and amygdaloidal, with associated masses of wacke: where the basalt is compact, it is either fine-grained or very coarsely crystallized; in the latter case it passes into an augitic rock, containing much olivine; the olivine is either colourless, or of the usual yellow and dull reddish shades. On some of the hills, beds of calcareous matter, both in an earthy and in a crystalline form, including

fragments of glossy scoriæ, are associated with the basaltic
strata. These strata differ from the streams of basaltic lava
forming the coast-plains, only in being more compact, and in
the crystals of augite, and in the grains of olivine being of
much greater size ;—characters which, together with the
appearance of the associated calcareous beds, induce me to
believe that they are of submarine formation.

Some considerable masses of wacke, which are associated
with these basaltic strata, and which likewise occur in the
basal series on the coast, especially at Quail Island, are
curious. They consist of a pale yellowish-green argillaceous
substance, of a crumbling texture when dry, but unctuous
when moist : in its purest form, it is of a beautiful green tint,
with translucent edges, and occasionally with obscure traces
of an original cleavage. Under the blowpipe it fuses very
readily into a dark gray, and sometimes even black bead,
which is slightly magnetic. From these characters, I naturally thought that it was one of the pale species decomposed, of the genus augite ;—a conclusion supported by the
unaltered rock being full of large separate crystals of black
augite, and of balls and irregular streaks of dark gray
augitic rock. As the basalt ordinarily consists of augite,
and of olivine often tarnished and of a dull red colour, I was
led to examine the stages of decomposition of this latter
mineral, and I found, to my surprise, that I could trace
a nearly perfect gradation from unaltered olivine to the
green wacke. Part of the same grain under the blowpipe
would in some instances behave like olivine, its colour being
only slightly changed, and part would give a black magnetic
bead. Hence I can have no doubt that the greenish wacke
originally existed as olivine; but great chemical changes
must have been effected during the act of decomposition,
thus to have altered a very hard, transparent, infusible
mineral, into a soft, unctuous, easily melted, argillaceous
substance.*

* D'Aubuisson, Traité de Géognosie (tom. ii. p. 569), mentions, on
the authority of M. Marcel de Serres, masses of green earth near Mont-

The basal strata of these hills, as well as some neighbouring, separate, bare, rounded hillocks, consist of compact, finegrained, non-crystalline (or so slightly as scarcely to be perceptible,) ferruginous feldspathic rocks, and generally in a state of semi-decomposition. Their fracture is exceedingly irregular, and splintery; yet small fragments are often very tough. They contain much ferruginous matter, either in the form of minute grains with a metallic lustre, or of brown hair-like threads; the rock in this latter case assuming a pseudo-brecciated structure. These rocks sometimes contain mica and veins of agate. Their rusty brown or yellowish colour is partly due to the oxides of iron, but chiefly to innumerable, microscopically minute, black specks, which, when a fragment is heated, are easily fused, and evidently are either hornblende or augite. These rocks, therefore, although at first appearing like baked clay or some altered sedimentary deposit, contain all the essential ingredients of trachyte; from which they differ only in not being harsh, and in not containing crystals of glassy feldspar. As is so often the case with trachytic formation, no stratification is here apparent. A person would not readily believe that these rocks could have flowed as lava; yet at St. Helena there are well characterized streams (as will be described in an ensuing chapter) of nearly similar composition. Amidst the hillocks composed of these rocks, I found in three places, smooth conical hills of phonolite,

péllier, which are supposed to be due to the decomposition of olivine. I do not, however, find, that the action of this mineral under the blowpipe being entirely altered, as it becomes decomposed, has been noticed; and the knowledge of this fact is important, as at first it appears highly improbable, that a hard, transparent, refractory mineral should be changed into a soft, easily-fused, clay, like this of St. Jago. I shall hereafter describe a green substance, forming threads within the cells of some vesicular basaltic rocks in Van Diemen's Land, which behave under the blowpipe like the green wacke of St. Jago; but its occurrence in cylindrical threads, shows it can not have resulted from the decomposition of olivine, a mineral always existing in the form of grains or crystals.

abounding with fine crystals of glassy feldspar, and with needles of hornblende. These cones of phonolite, I believe, bear the same relation to the surrounding feldspathic strata, which some masses of coarsely crystallized augitic rock, in another part of the island, bear to the surrounding basalt, namely, that both have been injected. The rocks of a feldspathic nature being anterior in origin to the basaltic strata, which cap them, as well as to the basaltic streams of the coast-plains, accords with the usual order of succession of these two grand divisions of the volcanic series.

The strata of most of these hills in the upper part, where alone the planes of division are distinguishable, are inclined at a small angle from the interior of the island towards the sea-coast. The inclination is not the same in each hill; in that marked A it is less than in B, D, or E; in C the strata are scarcely deflected from a horizontal plane, and in F (as far as I could judge without ascending it) they are slightly inclined in a reverse direction, that is, inwards and towards the centre of the island. Notwithstanding these differences of inclination, their correspondence in external form, and in the composition both of their upper and lower parts,—their relative position in one curved line, with their steepest sides turned inwards,—all seem to show that they originally formed parts of one platform; which platform, as before remarked, probably extended round a considerable portion of the circumference of the island. The upper strata certainly flowed as lava, and probably beneath the sea, as perhaps did the lower feldspathic masses : how then come these strata to hold their present position, and whence were they erupted ?

In the centre of the island * there are lofty mountains,

* I saw very little of the inland parts of the island. Near the village of St. Domingo, there are magnificent cliffs of rather coarsely crystallized basaltic lava. Following the little stream in this valley, about a mile above the village, the base of the great cliff was formed of a compact fine-grained basalt, conformably covered by a bed of pebbles. Near Fuentes, I met with pap-formed hills of the compact feldspathic series of rocks.

but they are separated from the steep inland flanks of these hills, by a wide space of lower country : the interior mountains, moreover, seem to have been the source of those great streams of basaltic lava, which, contracting as they pass between the bases of the hills in question, expand into the coast-plains. Round the shores of St. Helena there is a rudely-formed ring of basaltic rocks, and at Mauritius there are remnants of another such a ring round part, if not round the whole, of the island ; here again the same question immediately occurs, how come these masses to hold their present position, and whence were they erupted ? The same answer, whatever it may be, probably applies in these three cases ; and in a future chapter we shall recur to this subject.

Valleys near the coast.—These are broad, very flat, and generally bounded by low cliff-formed sides. Portions of the basaltic plain are sometimes nearly, or quite, isolated by them; of which fact, the space on which the town of Praya stands, offers an instance. The great valley west of the town, has its bottom filled up to a depth of more than twenty feet by well-rounded pebbles, which in some parts are firmly cemented together by white calcareous matter. There can be no doubt, from the form of these valleys, that they were scooped out by the waves of the sea, during that equable elevation of the land, of which the horizontal calcareous deposit, with its existing species of marine remains, gives evidence. Considering how well shells have been preserved in this stratum, it is singular that I could not find even a single small fragment of shell in the conglomerate at the bottom of the valleys. The bed of pebbles in the valley west of the town, is intersected by a second valley joining it as a tributary, but even this valley appears much too wide and flat-bottomed to have been formed by the small quantity of water, which falls only during one short wet season ; for at other times of the year, these valleys are absolutely dry.

Recent conglomerate.—On the shores of Quail Island, I found fragments of brick, bolts of iron, pebbles, and large

fragments of basalt, united by a scanty base of impure calcareous matter into a firm conglomerate. To show how
exceedingly firm this recent conglomerate is, I may mention,
that I endeavoured with a heavy geological hammer to
knock out a thick bolt of iron, which was embedded a little
above low-water mark, but was quite unable to succeed.

CHAPTER II.

FERNANDO NORONHA—*Precipitous hill of phonolite.*—TERCEIRA—*Trachytic rocks; their singular decomposition by steam of high temperature.*— TAHITI—*Passage from wacke into trap; singular volcanic rock with the vesicles half filled with mesotype.*—MAURITIUS—*Proofs of its recent elevation—Structure of its more ancient mountains; similarity with St. Jago.*— ST. PAUL'S ROCKS—*Not of volcanic origin—their singular mineralogical composition.*

Fernando Noronha.—During our short visit at this and the four following islands, I observed very little worthy of description. Fernando Noronha is situated in the Atlantic Ocean, in Lat. 3° 50′ S., and 230 miles distant from the coast of South America. It consists of several islets, together nine miles in length by three in breadth. The whole seems to be of volcanic origin; although there is no appearance of any crater, or of any one central eminence. The most remarkable feature is a hill 1000 feet high, of which the upper 400 feet consist of a precipitous, singularly-shaped pinnacle, formed of columnar phonolite, containing numerous crystals of glassy feldspar, and a few needles of hornblende. From the highest accessible point of this hill, I could distinguish in different parts of the group several other conical hills, apparently of the same nature. At St. Helena there are similar, great, conical, protuberant masses of phonolite, nearly 1000 feet in height, which have been formed by the injection of fluid feldspathic lava into yielding strata. If this hill has had, as is probable, a similar origin, denudation has been here effected on an enormous scale. Near the base of this hill, I observed beds of white tuff, intersected by numerous dikes, some of amygdaloidal basalt and others of trachyte; and beds of slaty phonolite with the planes of cleavage directed N. W. and S. E. Parts of this rock, where the crystals were scanty, closely resembled common clay-slate, altered by the contact of a trap-dike. The lamination of rocks, which undoubtedly have once been fluid, appears to me a subject well deserving attention. On the beach there were numerous fragments of compact basalt, of which rock, a distant façade of columns seemed to be formed.

Terceira in the Azores.—The central parts of this island consist of irregularly rounded mountains of no great elevation, composed of trachyte, which closely resembles in general character the trachyte of Ascension, presently to be described. This formation is in many parts overlaid, in the usual order of superposition, by streams of basaltic lava, which near the coast compose nearly the whole surface. The course which these streams have followed from their craters, can often be followed by the eye. The town of Angra is overlooked by a crateriform hill, (Mount Brazil) entirely built of thin strata of fine-grained, harsh, brown-coloured tuff. The upper beds are seen to overlap the basaltic streams, on which the town stands. This hill is almost identical in structure and composition with numerous crater-formed hills in the Galapagos Archipelago.

Effects of steam on the trachytic rocks.—In the central part of the island there is a spot, where steam is constantly issuing in jets from the bottom of a small ravine-like hollow, which has no exit, and which abuts against a range of trachytic mountains. The steam is emitted from several irregular fissures: it is scentless, soon blackens iron, and is of much too high a temperature to be endured by the hand. The manner in which the solid trachyte is changed on the borders of these orifices is curious: first, the base becomes earthy, with red freckles evidently due to the oxidation of particles of iron; then it becomes soft; and lastly, even the crystals of glassy feldspar yield to the dissolving agent. After the mass is converted into clay, the oxide of iron seems to be entirely removed from some parts, which are left perfectly white, whilst in other neighbouring parts, which are of the brightest red colour, it seems to be deposited in greater quantity; some other masses are marbled with the two distinct colours. Portions of the white clay, now that they are dry, cannot be distinguished by the eye from the finest prepared chalk; and when placed between the teeth they feel equally soft-grained; the inhabitants use this substance for white-washing their houses. The cause of the iron being dissolved in one part, and close by, being again deposited, is obscure; but the fact has been observed in several other places.* In some half-decayed specimens, I found small, globular, aggregations of yellow hyalite, resembling gumarabic, which no doubt had been deposited by the steam.

As there is no escape for the rain-water, which trickles down the sides of the ravine-like hollow, whence the steam issues, it must all percolate downwards through the fissures at its bottom. Some of the inhabitants informed me, that it was on record that flames (some luminous appearance?) had originally proceeded from these cracks, and that the flames had been succeeded by the steam; but I was not able to ascertain how long this was ago, or anything certain on the subject. When viewing the spot, I imagined that the injection of a large mass of rock, like the cone of phonolite at Fernando Noronha, in a semi-fluid state, by arching the surface might have caused a wedge-shaped hollow with cracks at the bottom, and that the rain-water percolating to the neighbourhood of the heated mass, would during many succeeding years be driven back in the form of steam.

Tahiti (Otaheite).—I visited only a part of the northwestern side of this island, and this part is entirely composed

---

* Spallanzani, Dolomieu and Hoffman have described similar cases in the Italian volcanic islands. Dolomieu says the iron at the Panza Islands is redeposited in the form of veins (p. 86, Mémoire sur les Isles Ponces). These authors likewise believe that the steam deposits silica: it is now experimentally known, that vapour of a high temperature is able to dissolve silica.

of volcanic rocks. Near the coast there are several varieties
of basalt, some abounding with large crystals of augite and
tarnished olivine, others compact and earthy,—some slightly
vesicular, and others occasionally amygdaloidal. These
rocks are generally much decomposed, and to my surprise, I
found in several sections, that it was impossible to distinguish,
even approximately, the line of separation between the
decayed lava and the alternating beds of tuff. Since the
specimens have become dry, it is rather more easy to distinguish the decomposed igneous rocks, from the sedimentary
tuffs. This gradation in character between rocks having
such widely different origins, may I think be explained by
the yielding under pressure of the softened sides of the vesicular cavities, which in many volcanic rocks occupy a large
proportion of their bulk. As the vesicles generally increase
in size and number in the upper parts of a stream of
lava, so would the effects of their compression increase ; the
yielding, moreover, of each lower vesicle must tend to disturb
all the softened matter above it. Hence we might expect to
trace a perfect gradation from an unaltered crystalline rock
to one in which all the particles (although originally forming
part of the same solid mass) had undergone mechanical displacement ; and such particles could hardly be distinguished
from others of similar composition, which had been deposited
as sediment. As lavas are sometimes laminated in their
upper parts, even horizontal lines, appearing like those of
aqueous deposition, could not in all cases be relied on as a
criterion of sedimentary origin. From these considerations
it is not surprising, that formerly many geologists believed in
real transitions from aqueous deposits, through wacke, into
igneous traps.

In the valley of Tia-auru, the commonest rocks are basalts
with much olivine, and in some cases almost composed of
large crystals of augite. I picked up some specimens, with
much glassy feldspar, approaching in character to trachyte.
There were also many large blocks of vesicular basalt, with
the cavities beautifully lined with chabasie (?), and radiating

bundles of mesotype. Some of these specimens presented a curious appearance, owing to a number of the vesicles being half filled up with a white, soft, earthy mesotypic mineral, which intumesced under the blow-pipe in a remarkable manner. As the upper surfaces in all the half-filled cells are exactly parallel, it is evident, that this substance has sunk to the bottom of each cell from its weight. Sometimes, however, it entirely fills the cells. Other cells are either quite filled, or lined, with small crystals, apparently of chabasie; these crystals, also, frequently line the upper half of the cells partly filled with the earthy mineral, as well as the upper surface of this substance itself, in which case the two minerals appear to blend into each other. I have never seen any other amygdaloid * with the cells half filled in the manner here described; and it is difficult to imagine the causes which determined the earthy mineral to sink from its gravity to the bottoms of the cells, and the crystalline mineral to adhere in a coating of equal thickness round the sides of the cells.

The basaltic strata on the sides of the valley, are gently inclined seaward, and I nowhere observed any sign of disturbance; the strata are separated from each other by thick, compact beds of conglomerate, in which the fragments are large, some being rounded, but most angular. From the character of these beds, from the compact and crystalline condition of most of the lavas, and from the nature of the infiltrated minerals, I was led to conjecture that they had originally flowed beneath the sea. This conclusion agrees with the fact, that the Rev. W. Ellis found marine remains at a considerable height, which he believes were interstratified with volcanic matter, as is likewise described to be the case by Messrs. Tyerman and Bennett at Huaheine, an island in

* MacCulloch, however, has described and given a plate of (Geolog. Trans. 1st Series, vol. iv. p. 225) a trap rock, with cavities filled up horizontally with quartz and chalcedony. The upper halves of these cavities are often filled by layers, which follow each irregularity of the surface, and by little depending stalactites of the same siliceous substances.

this same archipelago. Mr. Stutchbury also discovered near
the summit of one of the loftiest mountains of Tahiti, at the
height of several thousand feet, a stratum of semi-fossil coral.
None of these remains have been specifically examined. On
the coast, where masses of coral-rock would have afforded
the clearest evidence, I looked in vain for any signs of recent
elevation. For references to the above authorities, and for
more detailed reasons for not believing that Tahiti has been
recently elevated, I must refer to my volume (p. 138) on the
Structure and Distribution of Coral Reefs.

Mauritius.—Approaching this island on the northern or
north-western side, a curved chain of bold mountains, surmounted by rugged pinnacles, is seen to rise from a smooth
border of cultivated land, which gently slopes down to the
coast. At the first glance, one is tempted to believe that the
sea lately reached the base of these mountains, and upon
examination this view, at least with respect to the inferior
parts of the border, is found to be perfectly correct. Several
authors* have described masses of upraised coral-rock round
the greater part of the circumference of the island. Between
Tamarin Bay and the Great Black River, I observed, in company with Capt. Lloyd, two hillocks of coral-rock, formed in
their lower part of hard calcareous sandstone, and in their upper
of great blocks, slightly aggregated, of Astræa and Madrepora,
and of fragments of basalt; they were divided into beds dipping
seaward, in one case at an angle of 8°, and in the other at
18°; they had a water-worn appearance, and they rose
abruptly from a smooth surface, strewed with rolled debris of
organic remains, to a height of about twenty feet. The
Officier du Roi, in his most interesting tour in 1768 round
the island, has described masses of upraised coral-rocks, still

* Captain Carmichael, in Hooker's Bot. Misc. vol. ii. p. 301. Captain Lloyd has lately, in the Proceedings of the Geological Society
(vol. iii. p. 317), described carefully some of these masses. In the
"Voyage à l'Isle de France par un Officier du Roi," many interesting
facts are given on this subject. Consult also "Voyage aux Quatre Isles
d'Afrique, par M. Bory St. Vincent."

retaining that moat-like structure (p. 54 of my volume on Coral-Reefs) which is characteristic of the living reefs. On the coast northward of Port Louis, I found the lava concealed for a considerable space inland, by a conglomerate of corals and shells, like those on the beach, but in parts consolidated by red ferruginous matter. M. Bory St. Vincent has described similar calcareous beds over nearly the whole of the plain of Pamplemousses. Near Port Louis, when turning over some large stones, which lay in the bed of a stream at the head of a protected creek, and at the height of some yards above the level of spring-tides, I found several shells of serpula still adhering to their under sides.

The jagged mountains near Port Louis rise to a height of between 2000 and 3000 feet : they consist of strata of basalt, obscurely separated from each other by firmly aggregated beds of fragmentary matter ; and they are intersected by a few vertical dikes. The basalt in some parts abounds with large crystals of augite and olivine, and is generally compact. The interior of the island forms a plain, raised probably about a thousand feet above the level of the sea, and composed of streams of lava which have flowed round and between the rugged basaltic mountains. These more recent lavas are also basaltic, but less compact, and some of them abound with feldspar, so that they even fuse into a pale coloured glass. On the banks of the Great River, a section is exposed nearly 500 feet deep, worn through numerous thin sheets of the lava of this series, which are separated from each other by beds of scoriæ. They seem to have been of subaërial formation, and to have flowed from several points of eruption on the central platform, of which the Piton du Milieu is said to be the principal one. There are also several volcanic cones, apparently of this modern period, round the circumference of the island, especially at the northern end, where they form separate islets.

The mountains composed of the more compact and crystalline basalt, form the main skeleton of the island. M. Bailly*

* Voyage aux Terres Australes, tom. i. p. 54.

states that they all "se dévelopment autour d'elle comme
une ceinture d'immenses remparts, toutes afféctant une pente
plus ou moins inclinée vers le rivage de la mer, tandis au
contraire, que vers le centre de l'ile elles présentent une
coupe abrupte, et souvent taillée à pic. Toutes ces montagnes
sont formées de couches parallèles inclinées du centre de
l'île vers la mer." These statements have been disputed,
though not in detail, by M. Quoy, in the voyage of Freycinet.
As far as my limited means of observation went, I found
them perfectly correct.* The mountains on the N.W. side of
the island, which I examined, namely, La Pouce, Peter
Botts, Corps de Garde, Les Mamelles, and apparently
another further southward, have precisely the external shape
and stratification described by M. Bailly. They form about
a quarter of his girdle of ramparts. Although these mountains now stand quite detached, being separated from each
other by breaches, even several miles in width, through
which deluges of lava have flowed from the interior of the
island; nevertheless, seeing their close general similarity,
one must feel convinced that they originally formed parts of
one continuous mass. Judging from the beautiful map of
the Mauritius, published by the Admiralty from a French
MS., there is a range of mountains (M. Bamboo) on the
opposite side of the island, which correspond in height,
relative position, and external form with those just described.
Whether the girdle was ever complete may well be doubted;
but from M. Bailly's statements, and my own observations,
it may be safely concluded that mountains with precipitous
inland flanks, and composed of strata dipping outwards, once
extended round a considerable portion of the circumference
of the island. The ring appears to have been oval and of
vast size; its shorter axis, measured across from the inner
sides of the mountains near Port Louis and those near
Grand Port, being no less than thirteeen geographical miles
in length. M. Bailly boldly supposes that this enormous

* M. Lesson, in his account of this island, in the voyage of the Coquille, seems to follow M. Bailly's views.

gulf, which has since been filled up to a great extent by streams of modern lava, was formed by the sinking in of the whole upper part of one great volcano.

It is singular in how many respects those portions of St. Jago and of Mauritius which I visited, agree in their geological history. At both islands, mountains of similar external form, stratification, and (at least in their upper beds) composition, follow in a curved chain the coast-line. These mountains in each case appear originally to have formed parts of one continuous mass. The basaltic strata of which they are composed, from their compact and crystalline structure, seem, when contrasted with the neighbouring basaltic streams of subaërial formation, to have flowed beneath the pressure of the sea, and to have been subsequently elevated. We may suppose that the wide breaches between the mountains, were in both cases worn by the waves, during their gradual elevation,—of which process, within recent times, there is abundant evidence on the coast-land of both islands. At both, vast streams of more recent basaltic lavas have flowed from the interior of the island, round and between the ancient basaltic hills; at both, moreover, recent cones of eruption are scattered around the circumference of the island; but at neither have eruptions taken place within the period of history. As remarked in the last chapter, it is probable that these ancient basaltic mountains, which resemble (at least in many respects) the basal and disturbed remnants of two gigantic volcanos, owe their present form, structure, and position, to the action of similar causes.

St. Paul's Rocks.—This small island is situated in the Atlantic Ocean, nearly one degree north of the equator, and 540 miles distant from South America, in 29° 15′ west longitude. Its highest point is scarcely fifty feet above the level of the sea; its outline is irregular, and its entire circumference barely three-quarters of a mile. This little point of rock rises abruptly out of the ocean; and except on its western side, soundings were not obtained, even at the short distance of a quarter of a mile from its shore. It is

not of volcanic origin; and this circumstance, which is the
most remarkable point in its history (as will hereafter be
referred to), properly ought to exclude it from the present
volume. It is composed of rocks, unlike any which I have
met with, and which I cannot characterize by any name, and
must therefore describe.

The simplest, and one of the most abundant kinds, is
a very compact, heavy, greenish-black rock, having an
angular, irregular fracture, with some points just hard
enough to scratch glass, and infusible. This variety passes
into others of paler green tints, less hard, but with a more
crystalline fracture, and translucent on their edges; and
these are fusible into a green enamel. Several other varieties are chiefly characterized, by containing innumerable
threads of dark-green serpentine, and by having calcareous
matter in their interstices. These rocks have an obscure, concretionary structure, and are full of variously-coloured angular
pseudo fragments. These angular pseudo fragments consist
of the first-described dark green rock, of a brown softer kind,
of serpentine, and of a yellowish harsh stone, which, perhaps,
is related to serpentine rock. There are other vesicular,
calcareo-ferruginous, soft stones. There is no distinct
stratification, but parts are imperfectly laminated; and the
whole abounds with innumerable veins, and vein-like masses,
both small and large. Of these vein-like masses, some calcareous ones, which contain minute fragments of shells, are
clearly of subsequent origin to the others.

A glossy incrustation.—Extensive portions of these rocks
are coated by a layer of a glossy polished substance, with a
pearly lustre and of a grayish white colour; it follows all
the inequalities of the surface, to which it is firmly attached.
When examined with a lens, it is found to consist of numerous exceedingly thin layers, their aggregate thickness being
about the tenth of an inch. It is considerably harder than
calcareous spar, but can be scratched with a knife; under
the blow-pipe it scales off, decrepitates, slightly blackens,
emits a fetid odour, and becomes strongly alkaline: it does

not effervesce in acids.* I presume this substance has been deposited by water, draining from the birds' dung, with which the rocks are covered. At Ascension, near a cavity in the rocks, which was filled with a laminated mass of infiltrated birds' dung, I found some irregularly-formed, stalactitical masses of apparently the same nature. These masses when broken, had an earthy texture, but on their outsides, and especially at their extremities, they were formed of a pearly substance, generally in little globules, like the enamel of teeth, but more translucent, and so hard as just to scratch plate-glass. This substance slightly blackens under the blow-pipe, emits a bad smell, then becomes quite white, swelling a little, and fuses into a dull white enamel; it does not become alkaline; nor does it effervesce in acids. The whole mass had a collapsed appearance, as if in the formation of the hard glossy crust, the whole had shrunk much. At the Abrolhos Islands on the coast of Brazil, where also there is much birds' dung, I found a great quantity of a brown, arborescent substance adhering to some trap-rock. In its arboresent form, this substance singularly resembles some of the branched species of Nullipora. Under the blow-pipe, it behaves like the specimens from Ascension; but it is less hard and glossy, and the surface has not the shrunk appearance.

* In my Journal I have described this substance; I then believed that it was an impure phosphate of lime.

CHAPTER III.

THIS island is situated in the Atlantic ocean, in lat. 8° S.
long. 14° W.   It has the form of an irregular triangle, (see
accompanying map,) each side being about six miles in
length.   Its highest point is 2,870 feet* above the level of
the sea.   The whole is volcanic, and, from the absence of
proofs to the contrary, I believe of subaërial origin.   The
fundamental rock is everywhere of a pale colour, generally
compact, and of a feldspathic nature.   In the S. E. portion
of the island, where the highest land is situated, well characterized trachyte, and other congenerous rocks of that varying
family, occur.   Nearly the entire circumference is covered up
by black and rugged streams of basaltic lava, with here and
there a hill or single point of rock (one of which near the
sea-coast, north of the Fort, is only two or three yards across)
of the trachyte still remaining exposed.

Basaltic rocks. — The overlying basaltic lava is in some
parts extremely vesicular, in others little so ; it is of a
black colour, but sometimes contains crystals of glassy
feldspar, and seldom much olivine.   These streams appear

* Geographical Journal, vol. v. p. 243.

to have possessed singularly little fluidity; their side walls and lower ends being very steep, and even as much as between twenty and thirty feet in height. Their surface is extraordinarily rugged, and from a short distance appears as if studded with small craters. These projections consist of broad, irregularly conical, hillocks, traversed by fissures, and composed of the same unequally scoriaceous basalt with the surrounding streams, but having an obscure tendency to a columnar structure; they rise to a height between ten and thirty feet above the general surface, and have been formed, as I presume, by the heaping up of the viscid lava at points of greater resistance. At the base of several of these hillocks, and occasionally likewise on more level parts, solid ribs, composed of angulo-globular masses of basalt, resembling in size and outline arched sewers or gutters of brickwork, but not being hollow, project between two or three feet above the surface of the streams; what their origin may have been, I do not know. Many of the superficial fragments from these basaltic streams, present singularly convoluted forms; and some specimens could hardly be distinguished from logs of dark-coloured wood without their bark.

Many of the basaltic streams can be traced, either to points of eruption at the base of the great central mass of trachyte, or to separate, conical, red-coloured hills, which are scattered over the northern and western borders of the island. Standing on the central eminence, I counted between twenty and thirty of these cones of eruption. The greater number of them had their truncated summits cut off obliquely, and they all sloped towards the S.E., whence the trade-wind blows.* This structure no doubt has been caused, by the ejected fragments and ashes being always blown, during eruptions, in greater quantity towards one

---

* M. Lesson, in the Zoology of the Voyage of the Coquille (p. 490), has observed this fact. Mr. Hennah (Geolog. Proceedings, 1835, p. 189) further remarks, that the most extensive beds of ashes at Ascension invariably occur on the leeward side of the island.

side, than towards the other. M. Moreau de Jonnès has
made a similar observation with respect to the volcanic orifices in the West Indian islands.

Volcanic bombs.—These occur in great numbers strewed on
the ground, and some of them lie at considerable distances
from any points of eruption. They vary in size from that of
an apple, to that of a man's body ; they are either spherical
or pear-shaped, or with the hinder part (corresponding to
the tail of a comet), irregular, studded with projecting
points, and even concave. Their surfaces are rough and
fissured with branching cracks ; their internal structure is
either irregularly scoriaceous and compact, or it presents a
symmetrical and very curious appearance. An irregular
segment of a bomb, of this latter kind, of which I found
several, is accurately represented in the accompanying
woodcut. Its size was about that of a man's head.

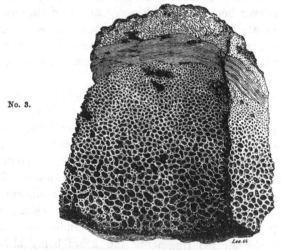

No. 3.

Fragment of a spherical volcanic bomb, with the interior parts coarsely cellular, coated by a
concentric layer of compact lava, and this again by a crust of finely cellular rock.

The whole interior is coarsely cellular ; the cells averaging
in diameter about the tenth of an inch ; but nearer the outside
they gradually decrease in size. This part is succeeded by a

well-defined shell of compact lava, having a nearly uniform thickness of about the third of an inch ; and the shell is overlaid by a somewhat thicker coating of finely cellular lava (the cells varying from the fiftieth to the hundredth of an inch in diameter), which forms the external surface: the line separating the shell of compact lava from the outer scoriaceous crust is distinctly defined. This structure is very simply explained, if we suppose a mass of viscid, scoriaceous matter, to be projected with a rapid, rotatory motion through the air; for whilst the external crust, from cooling, became solidified (in the state we now see it), the centrifugal force, by relieving the pressure in the interior parts of the bomb, would allow the heated vapours to expand their cells; but these being driven by the same force against the already-hardened crust, would become, the nearer they were to this part, smaller and smaller or less expanded, until they became packed into a solid, concentric shell. As we know that chips from a grindstone* can be flirted off, when made to revolve with sufficient velocity, we need not doubt that the centrifugal force would have power to modify the structure of a softened bomb, in the manner here supposed. Geologists have remarked, that the external form of a bomb at once bespeaks the history of its aërial course, and we now see that the internal structure can speak, with almost equal plainness, of its rotatory movement.

M. Bory St. Vincent† has described some balls of lava from the Isle of Bourbon, which have a closely similar structure; his explanation, however (if I understand it rightly), is very different from that which I have given ; for he supposes that they have rolled, like snow-balls, down the sides of the crater. M. Beudant,‡ also, has described some singular little balls of obsidian, never more than six or eight inches in diameter, which he found strewed on the surface of the ground: their form is always oval ; sometimes they are

* Nichol's Architecture of the Heavens.
† Voyage aux Quatre Isles d'Afrique, tom. i. p. 222.
‡ Voyage en Hongrie, tom. ii. p. 214.

much swollen in the middle, and even spindle-shaped: their
surface is regularly marked with concentric ridges and furrows, all of which on the same ball are at right angles to
one axis: their interior is compact and glassy. M. Beudant
supposes that masses of lava, when soft, were shot into the
air, with a rotatory movement round the same axis, and that
the form and superficial ridges of the bombs were thus produced. Sir Thomas Mitchell has given me what at first appears to be the half of a much flattened oval ball of obsidian;
it has a singular artificial-like appearance, which is well represented (of the natural size) in the accompanying woodcut.

No. 4.

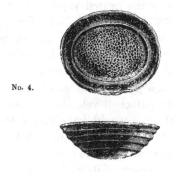

Volcanic bomb of obsidian from Australia. The upper figure gives a front view; the lower a
side view of the same object.

It was found in its present state, on a great sandy plain between the rivers Darling and Murray, in Australia, and at
the distance of several hundred miles from any known volcanic region. It seems to have been embedded in some
reddish tufaceous matter; and may have been transported
either by the aborigines or by natural means. The external
saucer consists of compact obsidian, of a bottle-green colour,
and is filled with finely-cellular black lava, much less transparent and glassy than the obsidian. The external surface
is marked with four or five, not quite, perfect ridges, which
are represented rather too distinctly in the woodcut. Here
then we have the external structure described by M. Beu-

dant, and the internal cellular condition of the bombs from
Ascension. The lip of the saucer is slightly concave, exactly
like the margin of a soup-plate, and its inner edge overlaps a
little the central, cellular lava. This structure is so symmetrical round the entire circumference, that one is forced to
suppose that the bomb burst during its rotatory course, before being quite solidified, and that the lip and edges were
thus slightly modified and turned inwards. It may be remarked that the superficial ridges are in planes, at right
angles to an axis, transverse to the longer axis of the flattened
oval: to explain this circumstance, we may suppose that
when the bomb burst, the axis of rotation changed.

Aëriform explosions.—The flanks of Green Mountain and
the surrounding country, are covered by a great mass, some
hundred feet in thickness, of loose fragments. The lower
beds generally consist of fine-grained, slightly consolidated
tuffs,* and the upper beds of great loose fragments, with alternating finer beds.† One white ribbon-like layer of decomposed, pumiceous breccia, was curiously bent into deep
unbroken curves, beneath each of the larger fragments in
the superincumbent stratum. From the relative position of
these beds, I presume that a narrow-mouthed crater, standing
nearly in the position of Green Mountain, like a great air-
gun, shot forth, before its final extinction, this vast accumulation of loose matter. Subsequently to this event, con-

---

* Some of this peperino, or tuff, is sufficiently hard not to be broken
by the greatest force of the fingers.

† On the northern side of the Green Mountain a thin seam, about
an inch in thickness, of compact oxide of iron extends over a considerable area; it lies conformably in the lower part of the stratified
mass of ashes and fragments. This substance is of a reddish-brown
colour, with an almost metallic lustre; it is not magnetic, but becomes
so, after having been heated under the blow-pipe, by which it is blackened and partly fused. This seam of compact stone, by intercepting
the little rain-water which falls on the island, gives rise to a small
dripping spring, first discovered by Dampier. It is the only fresh-water
on the island, so that the possibility of its being inhabited, has entirely
depended on the occurrence of this ferruginous layer.

siderable dislocations have taken place, and an oval circus
has been formed by subsidence. This sunken space lies at
the north-eastern foot of Green Mountain, and is well represented in the accompanying map. Its longer axis, which
is connected with a N. E. and S. W. line of fissure, is three-
fifths of a nautical mile in length; its sides are nearly
perpendicular, except in one spot, and about 400 feet in
height; they consist in the lower part, of a pale basalt with
feldspar, and in the upper part, of the tuff and loose ejected
fragments; the bottom is smooth and level, and under almost
any other climate, a deep lake would have been formed here.
From the thickness of the bed of loose fragments, with
which the surrounding country is covered, the amount of
aëriform matter, necessary for their projection, must have
been enormous; hence we may suppose it probable, that after
the explosions, vast subterranean caverns were left, and that
the falling in of the roof of one of these, produced the hollow
here described. At the Galapagos Archipelago, pits of a
similar character, but of a much smaller size, frequently
occur at the bases of small cones of eruption.

Ejected granitic fragments.—In the neighbourhood of Green
Mountain, fragments of extraneous rock are not unfrequently
found embedded in the midst of masses of scoriæ. Lieut.
Evans, to whose kindness I am indebted for much information, gave me several specimens, and I found others myself.
They nearly all have a granitic structure, are brittle, harsh
to the touch, and apparently of altered colours. *First,* a
white syenite, streaked and mottled with red; it consists of
well crystallized feldspar, numerous grains of quartz, and
brilliant, though small, crystals of hornblende. The feldspar
and hornblende in this and the succeeding cases, have been
determined by the reflecting goniometer, and the quartz by
its action under the blow-pipe. The feldspar in these ejected
fragments, like the glassy kind in the trachyte, is from its
cleavage a potash-feldspar. *Secondly*, a brick-red mass of
feldspar, quartz, and small dark patches of a decayed
mineral; one minute particle of which, I was able to ascer-

tain by its cleavage, to be hornblende. *Thirdly*, a mass of confusedly crystallized white feldspar, with little nests of a dark coloured mineral, often carious, externally rounded, having a glossy fracture, but no distinct cleavage: from comparison with the second specimen, I have no doubt that it is fused hornblende. *Fourthly*, a rock, which at first appears a simple aggregation of distinct and large-sized crystals of dusky-coloured Labrador feldspar;* but in their interstices there is some white granular feldspar, abundant scales of mica, a little altered hornblende, and, as I believe, no quartz. I have described these fragments in detail, because it is rare † to find granitic rocks ejected from volcanos with their *minerals unchanged*, as is the case with the first specimen, and partially with the second. One other large fragment, found in another spot, is deserving of notice; it is a conglomerate, containing small fragments of granitic, cellular, and jaspery rocks, and of hornstone porphyries, embedded in a base of wacke, threaded by numerous, thin

* Professor Miller has been so kind as to examine this mineral. He obtained two good cleavages of 86° 30′ and 86° 50′. The mean of several, which I made, was 86° 30′. Prof. Miller states that these crystals, when reduced to a fine powder, are soluble in hydrochloric acid, leaving some undissolved silex behind; the addition of oxalate of ammonia gives a copious precipitate of lime. He further remarks, that according to Von Kobell, anorthite (a mineral occurring in the ejected fragments at Mount Somna) is always white and transparent, so that if this be the case, these crystals from Ascension must be considered as Labrador feldspar. Prof. Miller adds, that he has seen an account, in " Erdmann's Journal für technische Chemie," of a mineral ejected from a volcano, which had the external characters of Labrador feldspar, but differed in the analysis, from that given by mineralogists, of this mineral: the author attributed this difference to an error in the analysis of Labrador feldspar, which is very old.

† Daubeny, in his work on Volcanos (p. 386), remarks that this is the case; and Humboldt, in his Personal Narrative (vol. i. p. 236), says, "In general, the masses of known primitive rocks, I mean those which perfectly resemble our granites, gneiss, and mica-slate, are very rare in lavas: the substances we generally denote by the name of granite, thrown out by Vesuvius, are mixtures of nepheline, mica, and pyroxene."

layers of a concretionary pitchstone passing into obsidian. These layers are parallel, slightly tortuous, and short; they thin out at their ends, and resemble in form, the layers of quartz in gneiss. It is probable that these small embedded fragments were not separately ejected, but were entangled in a fluid volcanic rock, allied to obsidian; and we shall presently see that several varieties of this latter series of rock assume a laminated structure.

Trachytic series of Rocks.—These occupy the more elevated and central, and likewise the south-eastern parts of the island. The trachyte is generally of a pale-brown colour, stained with small darker patches; it contains broken and bent crystals of glassy feldspar, grains of specular iron, and black microscopical points, which latter, from being easily fused, and then becoming magnetic, I presume are hornblende. The greater number of the hills, however, are composed of a quite white, friable stone, appearing like a trachytic tuff. Obsidian, hornstone, and several kinds of laminated feldspathic rocks, are associated with the trachyte. There is no distinct stratification; nor could I distinguish a crateriform structure in any of the hills of this series. Considerable dislocations have taken place; and many fissures in these rocks are yet left open, or are only partially filled with loose fragments. Within the space,* mainly formed of trachyte, some basaltic streams have burst forth; and not far from the summit of Green Mountain, there is one stream of quite black, vesicular basalt, containing minute crystals of glassy feldspar, which have a rounded appearance.

The soft white stone above-mentioned, is remarkable from its singular resemblance, when viewed in mass, to a sedimentary tuff: it was long before I could persuade myself that such was not its origin; and other geologists have been perplexed by closely similar formations in trachytic regions.

---

* This space is nearly included by a line sweeping round Green Mountain, and joining the hills, called the Weather Port Signal, Holyhead, and that denominated (improperly in a geological sense) "the crater of an old volcano."

In two cases, this white earthy stone formed isolated hills; in a third, it was associated with columnar and laminated trachyte; but I was unable to trace an actual junction. It contains numerous crystals of glassy feldspar and black microscopical specks, and is marked with small darker patches, exactly as in the surrounding trachyte. Its basis, however, when viewed under the microscope, is generally quite earthy; but sometimes it exhibits a decidedly crystalline structure. On the hill marked "Crater of an old volcano," it passes into a pale greenish-gray variety, differing only in its colour, and in not being so earthy; the passage was in one case effected insensibly; in another, it was formed by numerous, rounded and angular, masses of the greenish variety, being embedded in the white variety;—in this latter case, the appearance was very much like that of a sedimentary deposit, torn up and abraded, during the deposition of a subsequent stratum. Both these varieties are traversed by innumerable tortuous veins, (presently to be described,) which are totally unlike injected dikes, or indeed any other veins which I have ever seen. Both varieties include a few scattered fragments, large and small, of dark-coloured scoriaceous rocks, the cells of some of which are partially filled with the white earthy stone; they likewise include some huge blocks of a cellular porphyry.* These fragments project from the weathered surface, and perfectly resemble fragments embedded in a true sedimentary tuff. But as it is known, that extraneous fragments of cellular rock are sometimes included in columnar trachyte, in phonolite,† and in other compact lavas, this circumstance is not any real argument for the sedimentary origin of the white earthy stone.‡ The insensible passage of the greenish

---

* The porphyry is dark-coloured; it contains numerous, often fractured, crystals of white opaque feldspar, also decomposing crystals of oxide of iron; its vesicles include masses of delicate, hair-like, crystals, apparently of analcime.

† D'Aubuisson Traité de Géognosie, tom. ii. p. 548.

‡ Dr. Daubeny (on Volcanos, p. 180) seems to have been led to be-

variety into the white one, and likewise the more abrupt
passage by fragments of the former being embedded in
the latter, might result from slight differences in the composition of the same mass of molten stone, and from the abrading
action of one such part still fluid, on another part already
solidified.   The curiously formed veins have, I believe,
been formed by silicious matter being subsequently segre-
gated.   But my chief reason for believing that these soft
earthy stones, with their extraneous fragments, are not of
sedimentary origin, is the extreme improbability of crystals
of feldspar, black microscopical specks, and small stains of a
darker colour, occurring in the same proportional numbers
in an aqueous deposit, and in masses of solid trachyte.
Moreover, as I have remarked, the microscope occasionally
reveals a crystalline structure in the apparently earthy
basis.   On the other hand, the partial decomposition of such
great masses of trachyte, forming whole mountains, is
undoubtedly a circumstance of not easy explanation.

Veins in the earthy trachytic masses.—These veins are extraordinarily numerous, intersecting in the most complicated
manner both coloured varieties of the earthy trachyte: they
are best seen on the flanks of the " Crater of the old volcano."
They contain crystals of glassy feldspar, black microscopical
specks and little dark stains, precisely as in the surrounding
rock; but the basis is very different, being exceedingly
hard, compact, somewhat brittle, and of rather less easy fusibility.   The veins vary much, and suddenly, from the tenth of
an inch to one inch in thickness; they often thin out, not
only on their edges, but in their central parts, thus leaving
round, irregular apertures; their surfaces are rugged.

lieve that certain trachytic formations of Ischia and of the Puy de
Dôme, which closely resemble these of Ascension, were of sedimentary
origin, chiefly from the frequent presence in them "of scoriform portions, different in colour from the matrix." Dr. Daubeny adds, that on
the other hand, Brocchi, and other eminent geologists, have considered
these beds as earthy varieties of trachyte; he considers the subject
deserving of further attention.

They are inclined at every possible angle with the horizon, or are horizontal; they are generally curvilinear, and often interbranch one with another. From their hardness they withstand weathering, and projecting two or three feet above the ground, they occasionally extend some yards in length: these plate-like veins, when struck, emit a sound, almost like that of a drum, and they may be distinctly seen to vibrate; their fragments, which are strewed on the ground, clatter like pieces of iron, when knocked against each other. They often assume the most singular forms; I saw a pedestal of the earthy trachyte, covered by a hemispherical portion of a vein, like a great umbrella, sufficiently large to shelter two persons. I have never met with, or seen described, any veins like these; but in form they resemble the ferruginous seams, due to some process of segregation, occurring not uncommonly in sandstones,—for instance, in the New Red sandstone of England. Numerous veins of jasper and of siliceous sinter, occurring on the summit of this same hill, show that there has been some abundant source of silica, and as these plate-like veins differ from the trachyte, only in their greater hardness, brittleness, and less easy fusibility, it appears probable that their origin is due to the segregation or infiltration of siliceous matter, in the same manner as happens with the oxides of iron in many sedimentary rocks.

Siliceous sinter and jasper.—The siliceous sinter is either quite white, of little specific gravity, and with a somewhat pearly fracture, passing into pinkish pearly quartz; or it is yellowish white, with a harsh fracture, and it then contains an earthy powder in small cavities. Both varieties occur, either in large irregular masses in the altered trachyte, or in seams included in broad, vertical, tortuous, irregular veins of a compact, harsh, stone of a dull red colour, appearing like a sandstone. This stone, however, is only altered trachyte; and a nearly similar variety, but often honeycombed, sometimes adheres to the projecting plate-like veins, described in the last paragraph. The jasper

is of an ochre yellow or red colour; it occurs in large
irregular masses, and sometimes in veins, both in the altered
trachyte and in an associated mass of scoriaceous basalt.
The cells of the scoriaceous basalt are lined or filled with
fine, concentric layers of chalcedony, coated and studded
with bright-red oxide of iron. In this rock, especially in
the rather more compact parts, irregular angular patches
of the red jasper are included, the edges of which insensibly
blend into the surrounding mass ; other patches occur
having an intermediate character between perfect jasper
and the ferruginous, decomposed, basaltic base. In these
patches, and likewise in the large vein-like masses of jasper,
there occur little rounded cavities, of exactly the same
size and form with the air-cells, which in the scoriaceous
basalt are filled and lined with layers of chalcedony.
Small fragments of the jasper, examined under the microscope, seem to resemble the chalcedony with its colouring
matter not separated into layers, but mingled in the siliceous
paste, together with some impurities. I can understand
these facts,—namely, the blending of the jasper into the
semi-decomposed basalt,—its occurrence in angular patches,
which clearly do not occupy pre-existing hollows in the
rock,—and its containing little vesicles filled with chalcedony,
like those in the scoriaceous lava,—only on the supposi-
tion that a fluid, probably the same fluid which deposited
the chalcedony in the air-cells, removed in those parts
where there were no cavities, the ingredients of the basaltic
rock, and left in their place, silica and iron, and thus produced the jasper. In some specimens of silicified wood, I
have observed, that in the same manner as in the basalt, the
solid parts were converted into a dark-coloured homogeneous
stone, whereas the cavities formed by the larger sap-
vessels (which may be compared with the air-vesicles in the
basaltic lava) and other irregular hollows, apparently produced by decay, were filled with concentric layers of chalcedony; in this case, there can be little doubt that the
same fluid deposited the homogeneous base and the chal-

cedonic layers. After these considerations, I cannot doubt, but that the jasper of Ascension may be viewed as a volcanic rock silicified, in precisely the same sense as this term is applied to wood, when silicified: we are equally ignorant of the means by which every atom of wood, whilst in a perfect state, is removed and replaced by atoms of silica, as we are of the means by which the constituent parts of a volcanic rock could be thus acted on.* I was led to the careful examination of these rocks, and to the conclusion here given, from having heard the Rev. Professor Henslow express a similar opinion, regarding the origin in trap-rocks of many chalcedonies and agates. Siliceous deposits seem to be very general, if not of universal occurrence, in partially decomposed trachytic tuffs;† and as these hills, according to the view above given, consist of trachyte softened and altered *in situ*, the presence of free silica in this case, may be added as one more instance to the list.

Concretions in pumiceous tuff.—The hill, marked in the map " Crater of an old volcano," has no claims to this appellation, which I could discover, except in being surmounted by a circular, very shallow, saucer-like summit, nearly half a mile in diameter. This hollow has been nearly filled up with

---

* Beudant (Voyage en Hongrie, tom. iii. p. 502, 504) describes kidney-shaped masses of jasper-opal, which either blend into the surrounding trachytic conglomerate, or are embedded in it like chalk-flints; and he compares them with the fragments of opalized wood, which are abundant in this same formation. Beudant, however, appears to have viewed the process of their formation, rather as one of simple infiltration, than of molecular exchange; but the presence of a concretion, wholly different from the surrounding matter, if not formed in a pre-existing hollow, clearly seems to me to require, either a molecular or mechanical displacement of the atoms, which occupied the space afterwards filled by it. The jasper-opal of Hungary passes into chalcedony, and therefore in this case, as in that of Ascension, jasper seems to be intimately related in origin with chalcedony.

† Beudant (Voyage Min. tom. iii. p. 507) enumerates cases in Hungary, Germany, Central France, Italy, Greece, and Mexico.

many successive sheets of ashes and scoriæ, of different
colours, and slightly consolidated. Each successive, saucer-
shaped, layer crops out all round the margin, forming so
many rings of various colours, and giving to the hill a fan-
tastic appearance. The outer ring is broad, and of a white
colour; hence it resembles a course round which horses have
been exercised, and has received the name of the Devil's
Riding School, by which it is most generally known. These
successive layers of ashes must have fallen over the whole
surrounding country, but they have all been blown away
except in this one hollow, in which probably moisture accumulated, either during an extraordinary year when rain fell,
or during the storms often accompanying volcanic eruptions.
One of the layers of a pinkish colour, and chiefly derived
from small, decomposed fragments of pumice, is remarkable,
from containing numerous concretions. These are generally
spherical, from half-an-inch to three inches in diameter; but
they are occasionally cylindrical, like those of iron-pyrites in
the chalk of Europe. They consist of a very tough, compact, pale-brown stone, with a smooth and even fracture.
They are divided into concentric layers, by thin white par-
titions, resembling the external superficies; six or eight of
such layers are distinctly defined near the outside ; but those
toward the inside generally become indistinct, and blend into
a homogeneous mass. I presume that these concentric layers
were formed by the shrinking of the concretion, as it became
compact. The interior part is generally fissured by minute
cracks or septaria, which are lined, both by black, metallic, and
by other white and crystalline specks, the nature of which
I was unable to ascertain. Some of the larger concretions
consist of a mere spherical shell, filled with slightly consolidated ashes. The concretions contain a small proportion
of carbonate of lime : a fragment placed under the blow-
pipe decrepitates, then, whitens and fuses into a blebby
enamel, but does not become caustic. The surrounding
ashes do not contain any carbonate of lime ; hence the concretions have probably been formed, as is so often the case,

by the aggregation of this substance. I have not met with any account of similar concretions; and considering their great toughness and compactness, their occurrence in a bed, which probably has been subjected only to atmospheric moisture, is remarkable.

Formation of calcareous rocks on the sea-coast.—On several of the sea-beaches, there are immense accumulations of small, well-rounded particles of shells and corals, of white, yellowish, and pink colours, interspersed with a few volcanic particles. At the depth of a few feet, these are found cemented together into stone, of which the softer varieties are used for building; there are other varieties, both coarse and fine-grained, too hard for this purpose: and I saw one mass, divided into even layers half-an-inch in thickness, which were so compact, that when struck with a hammer they rang like flint. It is believed by the inhabitants, that the particles become united in the course of a single year. The union is effected by calcareous matter; and in the most compact varieties, each rounded particle of shell and volcanic rock can be distinctly seen to be enveloped in a husk of pellucid carbonate of lime. Extremely few perfect shells are embedded in these agglutinated masses; and I have examined even a large fragment under a microscope, without being able to discover the least vestige of striæ or other marks of external form: this shows how long each particle must have been rolled about, before its turn came to be embedded and cemented.* One of the most compact varieties, when placed in acid, was entirely dissolved, with the exception of some flocculent animal matter; its specific gravity was 2·63. The specific gravity of ordinary lime-stone varies from 2·6 to 2·75; pure Carrara marble was found by Sir H. De la Beche † to be 2·7. It is remarkable

---

* The eggs of the turtle being buried by the parent, sometimes become enclosed in the solid rock. Mr. Lyell has given a figure (Principles of Geology, book iii. ch. 17.) of some eggs, containing the bones of young turtles, found thus entombed.

† Researches in Theoretical Geology, p. 12.

that these rocks of Ascension, formed close to the surface, should be nearly as compact as marble, which has undergone the action of heat and pressure in the plutonic regions.

The great accumulation of loose calcareous particles, lying on the beach near the Settlement, commences in the month of October, moving towards the S.W., which, as I was informed by Lieut. Evans, is caused by a change in the prevailing direction of the currents. At this period the tidal rocks, at the S.W. end of the beach, where the calcareous sand is accumulating, and round which the currents sweep, become gradually coated with a calcareous incrustation, half-an-inch in thickness. It is quite white, compact, with some parts slightly spathose, and is firmly attached to the rock. After a short time it gradually disappears, being either redissolved, when the water is less charged with lime, or more probably is mechanically abraded. Lieut. Evans has observed these facts, during the six years he has resided at Ascension. The incrustation varies in thickness in different years : in 1831 it was unusually thick. When I was there in July, there was no remnant of the incrustation ; but on a point of basalt, from which the quarrymen had lately removed a mass of the calcareous freestone, the incrustation was perfectly preserved. Considering the position of the tidal rocks, and the period at which they become coated, there can be no doubt, that the movement and disturbance of the vast accumulation of calcareous particles, many of them being partially agglutinated together, causes the waves of the sea to be so highly charged with carbonate of lime, that they deposit it on the first objects against which they impinge. I have been informed by Lieut. Holland, R.N., that this incrustation is formed on many parts of the coast, on most of which, I believe, there are likewise great masses of comminuted shells.

A frondescent calcareous incrustation.—In many respects this is a singular deposit ; it coats throughout the year the tidal volcanic rocks, that project from the beaches composed of broken shells. Its general appearance is well re-

presented in the accompanying woodcut; but the fronds or
discs, of which it is composed, are generally so closely

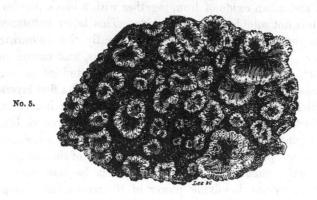

No. 5.

An Incrustation of calcareous and animal matter, coating the tidal rocks at Ascension.

crowded together as to touch. These fronds have their
sinuous edges finely crenulated, and they project over their
pedestals or supports; their upper surfaces are either slightly
concave, or slightly convex; they are highly polished, and of
a dark gray or jet black colour; their form is irregular,
generally circular, and from the tenth of an inch to one inch
and a-half in diameter; their thickness, or amount of their
projection from the rock on which they stand, varies much,
about a quarter of an inch being perhaps most usual. The
fronds occasionally become more and more convex, until
they pass into botryoidal masses with their summits fissured;
when in this state, they are glossy and of an intense black,
so as to resemble some fused metallic substance. I have
shown the incrustation, both in this latter and in its ordinary
state, to several geologists, but not one could conjecture its
origin, except that perhaps it was of volcanic nature!

The substance forming the fronds, has a very compact and
often almost crystalline fracture; the edges being translucent, and hard enough easily to scratch calcareous spar.
Under the blow-pipe it immediately becomes white, and
emits a strong animal odour, like that from fresh shells. It

is chiefly composed of carbonate of lime; when placed in
muriatic acid it froths much, leaving a residue of sulphate of
lime, and of an oxide of iron, together with a black powder,
which is not soluble in heated acids.   This latter substance
seems to be carbonaceous, and is evidently the colouring
matter.   The sulphate of lime is extraneous, and occurs in
distinct, excessively minute, lamellar plates, studded on the
surfaces of the fronds, and embedded between the fine layers
of which they are composed; when a fragment is heated in
the blowpipe, these lamellæ are immediately rendered visible.
The original outline of the fronds may often be traced, either
to a minute particle of shell fixed in a crevice of the rock, or
to several cemented together; these first become deeply
corroded, by the dissolving power of the waves, into sharp
ridges, and then are coated with successive layers of the
glossy, gray, calcareous incrustation.   The inequalities of the
primary support affect the outline of every successive layer,
in the same manner as may often be seen in bezoar-stones,
when an object like a nail forms the centre of aggregation.
The crenulated edges, however, of the frond appear to be
due to the corroding power of the surf on its own deposit,
alternating with fresh depositions.   On some smooth basaltic
rocks on the coast of St. Jago, I found an exceedingly thin layer
of brown calcareous matter, which under a lens presented a
miniature likeness of the crenulated and polished fronds of
Ascension; in this case a basis was not afforded by any projecting extraneous particles.   Although the incrustation at
Ascension, is persistent throughout the year; yet from the
abraded appearance of some parts, and from the fresh appearance of other parts, the whole seems to undergo a round
of decay and renovation, due probably to changes in the form
of the shifting beach, and consequently in the action of the
breakers: hence probably it is, that the incrustation never
acquires a great thickness.   Considering the position of the
encrusted rocks in the midst of the calcareous beach, together with its composition, I think there can be no doubt,
that its origin is due to the dissolution and subsequent depo-

sition of the matter, composing the rounded particles of
shells and corals.* From this source it derives its animal
matter, which is evidently the colouring principle. The
nature of the deposit, in its incipient stage, can often be well
seen upon a fragment of white shell, when jammed between
two of the fronds; it then appears exactly like the thinnest
wash of a pale gray varnish. Its darkness varies a little,
but the jet blackness of some of the fronds and of the botryoidal masses, seems due to the translucency of the successive
gray layers. There is, however, this singular circumstance,
that when deposited on the under side of ledges of rock or
in fissures, it appears always to be of a pale, pearly gray
colour, even when of considerable thickness: hence one is
led to suppose, that an abundance of light is necessary to the
development of the dark colour, in the same manner as
seems to be the case with the upper and exposed surfaces of the
shells of living mollusca, which are always dark, compared
with their under surfaces and with the parts habitually covered
by the mantle of the animal. In this circumstance,—in the
immediate loss of colour and in the odour emitted under the
blow-pipe,—in the degree of hardness and translucency of
the edges,—and in the beautiful polish of the surface,† rival-

* The selenite, as I have remarked, is extraneous, and must have
been derived from the sea-water. It is an interesting circumstance
thus to find the waves of the ocean, sufficiently charged with sulphate
of lime, to deposit it on the rocks, against which they dash every tide.
Dr. Webster has described (Voyage of the Chanticleer, vol. ii. p. 319)
beds of gypsum and salt, as much as two feet in thickness, left by the
evaporation of the spray on the rocks on the windward coast. Beautiful
stalactites of selenite, resembling in form those of carbonate of lime,
are formed near these beds. Amorphous masses of gypsum, also, occur in
caverns in the interior of the island; and at Cross Hill (an old crater)
I saw a considerable quantity of salt oozing from a pile of scoriæ. In
these latter cases, the salt and gypsum appear to be volcanic products.
† From the fact described in my Journal of Researches (p. 12), of a
coating of oxide of iron, deposited by a streamlet on the rocks in its bed
(like a nearly similar coating at the great cataracts of the Orinooco and
Nile), becoming finely polished where the surf acts, I presume that
the surf in this instance, also, is the polishing agent.

ling when in a fresh state that of the finest Oliva, there is a
striking analogy between this inorganic incrustation and the
shells of living molluscous animals.* This appears to me
to be an interesting physiological fact.†

*Singular laminated beds alternating with and passing into
obsidian.*—These beds occur within the trachytic district, at
the western base of Green Mountain, under which they dip
at a high inclination. They are only partially exposed,
being covered up by modern ejections; from this cause,
I was unable to trace their junction with the trachyte, or to
discover whether they had flowed as a stream of lava, or
had been injected amidst the overlying strata. There are
three principal beds of obsidian, of which the thickest forms
the base of the section. The alternating stony layers appear
to me eminently curious, and shall be first described, and
afterwards their passage into the obsidian. They have an
extremely diversified appearance; five principal varieties
may be noticed, but these insensibly blend into each other
by endless gradations.

First,—A pale gray, irregularly and coarsely laminated,‡

* In the section descriptive of St. Paul's Rocks, I have described a
glossy, pearly substance, which coats the rocks, and an allied stalactitical incrustation from Ascension, the crust of which resembles the
enamel of teeth, but is hard enough to scratch plate glass. Both these
substances contain animal matter, and seem to have been derived from
water infiltering through birds' dung.

† Mr. Horner and Sir David Brewster have described (Philosophical
Transactions, 1836, p. 65) a singular "artificial substance, resembling
shell." It is deposited in fine, transparent, highly polished, brown-
coloured laminæ, possessing peculiar optical properties, on the inside of
a vessel, in which cloth, first prepared with glue and then with lime, is
made to revolve rapidly in water. It is much softer, more transparent,
and contains more animal matter, than the natural incrustation at
Ascension; but we here again see, the strong tendency which carbonate
of lime and animal matter evince to form a solid substance allied to
shell.

‡ This term is open to some misinterpretation, as it may be applied
both to rocks divided into laminæ of exactly the same composition,
and to layers firmly attached to each other, with no fissile tendency, but

harsh-feeling rock, resembling clay-slate which has been in contact with a trap-dike, and with a fracture of about the same degree of crystalline structure. This rock, as well as the following varieties, easily fuse into a pale glass. The greater part is honey-combed with irregular, angular, cavities, so that the whole has a carious appearance, and some fragments resemble in a remarkable manner silicified logs of decayed wood. This variety, especially where more compact, is often marked with thin whitish streaks, which are either straight or wrap round, one behind the other, the elongated carious hollows.

Secondly,—A bluish gray or pale brown, compact, heavy, homogeneous stone, with an angular, uneven, earthy fracture; viewed, however, under a lens of high power, the fracture is seen to be distinctly crystalline, and even separate minerals can be distinguished.

Thirdly,—A stone of the same kind with the last, but streaked with numerous, parallel, slightly tortuous, white lines of the thickness of hairs. These white lines are more crystalline than the parts between them; and the stone splits along them: they frequently expand into exceedingly thin cavities, which are often only just perceptible with a lens. The matter forming the white lines becomes better crystallized in these cavities, and Prof. Miller was fortunate enough, after several trials, to ascertain that the white crystals, which are the largest, were of quartz,* and that the minute green transparent needles were augite, or, as they would more generally be called, diopside: besides these crystals, there are some minute, dark specks without a trace of crystallization, and some fine, white, granular, crys-

composed of different minerals, or of different shades of colour. The term laminated, in this chapter, is applied in these latter senses; where a homogeneous rock splits, as in the former sense, in a given direction, like clay-slate, I have used the term fissile.

* Professor Miller informs me __ that the crystals which he measured had the faces *P*, *z*, *m* of the figure (147) given by Haidinger in his Translation of Mohs; and he adds, that it is remarkable, that none of them had the slightest trace of faces *r* of the regular six-sided prism.

talline matter, which is probably feldspar. Minute fragments
of this rock are easily fusible.

Fourthly,—A compact crystalline rock, banded in straight
lines with innumerable layers of white and gray shades of
colour, varying in width from the $\frac{1}{30}$th to the $\frac{1}{200}$th of an
inch ; these layers seem to be composed chiefly of feldspar,
and they contain numerous perfect crystals of glassy feldspar, which are placed lengthways ; they are also thickly
studded with microscopically minute, amorphous, black
specks, which are placed in rows, either standing separately,
or more frequently united, two or three or several together,
into black lines, thinner than a hair. When a small fragment
is heated in the blow-pipe, the black specks are easily fused
into black brilliant beads, which become magnetic,—characters that apply to no common mineral except hornblende or
augite. With the black specks, there are mingled some
others of a red colour, which are magnetic before being
heated, and no doubt are oxide of iron. Round two little
cavities, in a specimen of this variety, I found the black
specks aggregated into minute crystals, appearing like those
of augite or hornblende, but too dull and small to be measured by the goniometer; in this specimen, also, I could
distinguish amidst the crystalline feldspar, grains, which had
the aspect of quartz. By trying with a parallel ruler,
I found that the thin gray layers, and the black hair-like lines,
were absolutely straight and parallel to each other. It is
impossible to trace the gradation from the homogeneous gray
rocks to these striped varieties, or indeed the character of
the different layers in the same specimen, without feeling
convinced that the more or less perfect whiteness of the
crystalline feldspathic matter, depends on the more or less
perfect aggregation of diffused matter, into the black and
red specks of hornblende and oxide of iron.

Fifthly,—A compact heavy rock, not laminated, with an
irregular, angular, highly crystalline, fracture ; it abounds
with distinct crystals of glassy feldspar, and the crystalline
feldspathic base is mottled with a black mineral, which on

the weathered surface is seen to be aggregated into small crystals, some perfect, but the greater number imperfect. I showed this specimen to an experienced geologist, and asked him what it was, he answered, as I think every one else would have done, that it was a primitive greenstone. The weathered surface, also, of the foregoing (No. 4) banded variety, strikingly resembles a worn fragment of finely laminated gneiss.

These five varieties, with many intermediate ones, pass and repass into each other. As the compact varieties are quite subordinate to the others, the whole may be considered as laminated or striped. The laminæ, to sum up their characteristics, are either quite straight, or slightly tortuous, or convoluted; they are all parallel to each other, and to the intercalating strata of obsidian; they are generally of extreme thinness; they consist either of an apparently homogeneous, compact rock, striped with different shades of · gray and brown colours, or of crystalline feldspathic layers in a more or less perfect state of purity, and of different thicknesses, with distinct crystals of glassy feldspar placed lengthways, or of very thin layers chiefly composed of minute crystals of quartz and augite, or composed of black and red specks of an augitic mineral and of an oxide of iron, either not crystallized or imperfectly so. After having fully described the obsidian, I shall return to the subject of the lamination of rocks of the trachytic series.

The passage of the foregoing beds into the strata of glassy obsidian is effected in several ways : first, angulo-modular masses of obsidian, both large and small, abruptly appear disseminated in a slaty, or in an amorphous, pale-coloured feldspathic rock, with a somewhat pearly fracture. Secondly, small irregular nodules of the obsidian, either standing separately, or united into thin layers, seldom more than the tenth of an inch in thickness, alternate repeatedly with very thin layers of a feldspathic rock, which is striped with the finest parallel zones of colour, like an agate, and which sometimes passes into the nature of pitchstone; the interstices

between the nodules of obsidian are generally filled by soft
white matter, resembling pumiceous ashes. Thirdly, the
whole substance of the bounding rock suddenly passes into
an angulo-concretionary mass of obsidian. Such masses (as
well as the small nodules) of obsidian are of a pale green
colour, and are generally streaked with different shades of
colour, parallel to the laminæ of the surrounding rock; they
likewise generally contain minute white sphærulites, of which
half is sometimes embedded in a zone of one shade of colour,
and half in a zone of another shade. The obsidian assumes
its jet black colour and perfectly conchoidal fracture, only
when in large masses; but even in these, on careful examination and on holding the specimens in different lights, I
could generally distinguish parallel streaks of different shades
of darkness.

One of the commonest transitional rocks deserves in
several respects a further description. It is of a very complicated nature, and consists of numerous thin, slightly
tortuous, layers of a pale-coloured feldspathic stone, often
passing into an imperfect pitchstone, alternating with layers
formed of numberless little globules of two varieties of obsidian, and of two kinds of sphærulites, embedded in a soft or
in a hard pearly base. The sphærulites are either white and
translucent, or dark brown and opaque; the former are
quite spherical, of small size, and distinctly radiated from
their centre. The dark brown sphærulites are less perfectly
round, and vary in diameter from the $\frac{1}{20}$ to $\frac{1}{30}$ of an inch;
when broken they exhibit towards their centres, which are
whitish, an obscure radiating structure; two of them when
united, sometimes have only one central point of radiation;
there is occasionally a trace of a hollow or crevice in their
centres. They stand either separately, or are united two or
three or many together into irregular groups, or more
commonly into layers, parallel to the stratification of the
mass. This union in many cases is so perfect, that the two
sides of the layer thus formed, are quite even; and these
layers, as they become less brown and opaque, cannot be

distinguished from the alternating layers of the pale-coloured feldspathic stone. The sphærulites, when not united, are generally compressed in the plane of the lamination of the mass; and in this same plane, they are often marked internally, by zones of different shades of colour, and externally by small ridges and furrows. In the upper part of the accompanying woodcut, the sphærulites with the parallel ridges and furrows

No. 6.

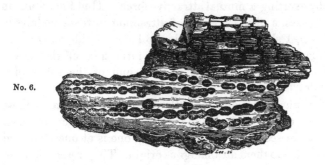

Opaque brown sphærulites, drawn on an enlarged scale. The upper ones are externally marked with parallel ridges. The internal radiating structure of the lower ones, is much too plainly represented.

are represented on an enlarged scale, but they are not well executed; and in the lower part, their usual manner of grouping is shown. In another specimen, a thin layer formed of the brown sphærulites closely united together, intersects, as represented in the woodcut, No. 7, a layer of

No. 7.

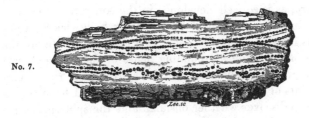

A layer formed by the union of minute brown sphærulites, intersecting two other similar layers : the whole represented of nearly the natural size.

similar composition; and after running for a short space in a slightly curved line, again intersects it, and likewise a second layer lying a little way beneath that first intersected.

The small nodules also of obsidian are sometimes externally marked with ridges and furrows, parallel to the lamination of the mass, but always less plainly than the sphærulites. These obsidian nodules are generally angular, with their edges blunted; they are often impressed with the form of the adjoining sphærulites, than which they are always larger; the separate nodules seldom appear to have drawn each other out by exerting a mutual attractive force. Had I not found in some cases, a distinct centre of attraction in these nodules of obsidian, I should have been led to have considered them as residuary matter, left during the formation of the pearl-stone, in which they are embedded, and of the sphærulitic globules.

The sphærulites and the little nodules of obsidian in these rocks, so closely resemble in general form and structure, con-cretions in sedimentary deposits, that one is at once tempted to attribute to them an analogous origin. They resemble ordi-nary concretions in the following respects,—in their external form—in the union of two or three, or of several, into an irre-gular mass, or into an even-sided layer,—in the occasional in-tersection of one such layer by another, as in the case of chalk-flints,—in the presence of two or three kinds of nodules, often close together, in the same basis,—in their fibrous, radi-ating structure, with occasional hollows in their centres,—in the co-existence of a laminary, concretionary, and radiating structure, as is so well developed in the concretions of mag-nesian limestone, described by Professor Sedgwick.* Con-cretions in sedimentary deposits, it is known, are due to the separation from the surrounding mass of the whole or part of some mineral substance, and its aggregation round cer-tain points of attraction. Guided by this fact, I have endea-voured to discover whether obsidian and the sphærulites (to which may be added marekanite and pearlstone, both of them occurring in nodular concretions in the trachytic series) differ in their constituent parts, from the minerals generally composing trachytic rocks. It appears from three

* Geological Transactions, vol. iii. part i. p. 37.

analyses, that obsidian contains on an average 76 per cent. of silica; from one analysis, that sphærulites contain 79·12; from two, that marekanite contains 79·25; and from two other analyses, that pearlstone contains 75·62 of silica.* Now, the constituent parts of trachyte, as far as they can be distinguished, consist of feldspar, containing 65·21 of silica; or of albite containing 69·09; of hornblende containing 55·27,† and of oxide of iron: so that the foregoing glassy concretionary substance, all contain a larger proportion of silica than that occurring in ordinary feldspathic or trachytic rocks. D'Aubuisson,‡ also, has remarked on the large proportion of silica compared with alumina, in six analyses of obsidian and pearlstone given in Brongniart's Mineralogy. Hence I conclude, that the foregoing concretions have been formed by a process of aggregation, strictly analogous to that which takes place in aqueous deposits, acting chiefly on the silica, but likewise on some of the other elements of the surrounding mass, and thus producing the different concretionary varieties. From the well-known effects of rapid cooling§ in giving glassiness of texture, it is probably necessary that the entire mass, in cases like that of Ascension, should have cooled at a certain rate; but considering the repeated and complicated alternations, of nodules and thin layers of a glassy texture with other layers quite stony or crystalline, all within the space of a few feet or even inches, it is hardly possible that they could have cooled at different rates, and thus have acquired their different textures.

The natural sphærulites in these rocks,‖ very closely

* The foregoing analyses are taken from Beudant Traité de Minéralogie, tom. ii. p. 113; and one analysis of obsidian, from Phillips's Mineralogy.

† These analyses are taken from Von Kobell's, Grundzüge der Mineralogie, 1838.

‡ Traité de Géogn. tom. ii. p. 535.

§ This is seen in the manufactory of common glass, and in Gregory Watts's experiments on molten trap; also on the natural surfaces of lava-streams, and on the side-walls of dikes.

‖ I do not know whether it is generally known, that bodies having

resemble those produced in glass, when slowly cooled. In some fine specimens of partially devitrified glass, in the possession of Mr. Stokes, the sphærulites are united into straight layers with even sides, parallel to each other, and to one of the outer surfaces, exactly as in the obsidian. These layers sometimes interbranch and form loops; but I did not see any case of actual intersection. They form the passage from the perfectly glassy portions, to those nearly homogeneous and stony, with only an obscure concretionary structure. In the same specimen, also, sphærulites differing slightly in colour and in structure, occur embedded close together. Considering these facts, it is some confirmation of the view above given of the concretionary origin of the obsidian and natural sphærulites, to find that M. Dartigues,* in his curious paper on this subject, attributes the production of sphærulites in glass, to the different ingredients obeying their own laws of attraction and becoming aggregated. He is led to believe that this takes place, from the difficulty in remelting sphærulitic glass, without the whole be first thoroughly pounded and mixed together; and likewise from the fact, that the change takes place most readily in glass composed of many ingredients. In confirmation of M. Dartigues' view, I may remark, that M. Fleuriau de Bellevue† found that the sphærulitic portions of devitrified glass were acted on both by nitric acid and under the blow-pipe, in a different manner from the compact paste in which they were embedded.

exactly the same appearance as sphærulites, sometimes occur in agates. Mr. Robert Brown showed me in an agate, formed within a cavity in a piece of silicified wood, some little specks, which were only just visible to the naked eye: these specks, when placed by him under a lens of high power, presented a beautiful appearance: they were perfectly circular, and consisted of the finest fibres of a brown colour, radiating with great exactness from a common centre. These little radiating stars are occasionally intersected, and portions are quite cut off by the fine, ribbon-like zones of colour in the agate. In the obsidian of Ascension, the halves of a sphærulite often lie in different zones of colour, but they are not cut off by them, as in the agate.

* Journal de Physique, tom. 59 (1804), pp. 10, 12.

† Idem, tom. 60 (1805), p. 418.

Comparison of the obsidian beds and alternating strata of Ascension, with those of other countries.—I have been struck with much surprise, how closely the excellent description of the obsidian rocks of Hungary, given by Beudant,* and that by Humboldt, of the same formation in Mexico and Peru,† and likewise the descriptions given by several authors‡ of the trachytic regions in the Italian islands, agree with my observations at Ascension. Many passages might have been transferred without alteration from the works of the above authors, and would have been applicable to this island. They all agree in the laminated and stratified character of the whole series; and Humboldt speaks of some of the beds of obsidian, being ribboned like jasper.§ They all agree in the nodular or concretionary character of the obsidian, and of the passage of these nodules into layers. They all refer to the repeated alternations, often in undulatory planes, of glassy, pearly, stony, and crystalline layers : the crystalline layers, however, seem to be much more perfectly developed

* Voyage en Hongrie, tom. i. p. 330; tom. ii. pp. 221 & 315; tom. iii. pp. 369, 371, 377, 381.

† Essai Géognostique, pp. 176, 326, 328.

‡ P. Scrope, in Geological Transactions, vol. ii. (second series) p. 195. Consult, also, Dolimieu's Voyage aux isles Lipari, and D'Aubuisson Traité de Géogn. tom. ii. p. 534.

§ In Mr Stokes' fine collection of obsidians from Mexico, I observe that the sphærulites are generally much larger than those of Ascension; they are generally white, opaque, and are united into distinct layers : there are many singular varieties, different from any at Ascension. The obsidians are finely zoned, in quite straight or curved lines, with exceedingly slight differences of tint, of cellularity, and of more or less perfect degrees of glassiness. Tracing some of the less perfectly glassy zones, they are seen to become studded with minute white sphærulites, which become more and more numerous, until at last they unite and form a distinct layer : on the other hand, at Ascension, only the brown sphærulites unite and form layers; the white ones always being irregularly disseminated. Some specimens at the Geological Society, said to belong to an obsidian formation from Mexico, have an earthy fracture, and are divided in the finest parallel laminæ, by specks of a black mineral, like the augitic or hornblendic specks in the rocks at Ascension.

at Ascension, than in the above-named countries. Humboldt compares some of the stony beds, when viewed from a distance, to strata of a schistose sandstone. Sphærulites are described as occurring abundantly in all cases; and they everywhere seem to mark the passage, from the perfectly glassy to the stony and crystalline beds. Beudant's account* of his " perlite lithoide globulaire" in every, even the most trifling particular, might have been written for the little brown sphærulitic globules of the rocks of Ascension.

From the close similarity in so many respects, between the obsidian formations of Hungary, Mexico, Peru, and of some of the Italian islands, with that of Ascension, I can hardly doubt that in all these cases, the obsidian and the sphærulites owe their origin to a concretionary aggregation of the silica, and of some of the other constituent elements, taking place whilst the liquified mass cooled at a certain required rate. It is, however, well known, that in several places, obsidian has flowed in streams like lava; for instance, at Teneriffe, at the Lipari islands, and at Iceland.† In these cases, the superficial parts are the most perfectly glassy, the obsidian passing at the depth of a few feet into an opaque stone. In an analysis by Vauquelin of a specimen of obsidian from Hecla, which probably flowed as lava, the proportion of silica is nearly the same as in the nodular or concretionary obsidian from Mexico. It would be interesting to ascertain, whether the opaque interior portions and the superficial glassy coating, contained the same proportional constituent parts: we know from M. Dufrénoy‡ that the exterior and interior parts of the same stream of lava, sometimes differ considerably in their composition. Even should the whole body of the stream of obsidian turn out to be

---

* Beudant's Voyage, tom. iii. p. 373.

† For Teneriffe, see Von Buch Descript. des isles Canaries, p. 184 and 190; for the Lipari Islands, see Dolimieu's Voyage, p. 34; for Iceland, see Mackenzie's Travels, p. 369.

‡ Mémoires pour servir a une descript. Géolog. de la France, tom. iv. p. 371.

similarly composed with nodular obsidian, it would only be necessary, in accordance with the foregoing facts, to suppose that lava in these instances had been erupted with its ingredients mixed in the same proportion, as in the concretionary obsidian.

Lamination of volcanic rocks of the trachytic series.

We have seen, that in several and widely distant countries, the strata alternating with beds of obsidian, are highly laminated. The nodules, also, both large and small, of the obsidian, are zoned with different shades of colour; and I have seen a specimen from Mexico in Mr. Stokes' collection, with its external surface weathered* into ridges and furrows, corresponding with the zones of different degrees of glassiness: Humboldt,† moreover, found on the Peak of Teneriffe, a stream of obsidian divided by very thin, alternating, layers of pumice. Many other lavas of the feldspathic series are laminated; thus, masses of common trachyte at Ascension, are divided by fine earthy lines, along which the rock splits, separating thin layers of slightly different shades of colour; the greater number, also, of the embedded crystals of glassy feldspar are placed lengthways in the same direction. Mr. P. Scrope‡ has described a remarkable columnar trachyte in the Panza Islands, which seems to have been injected into an overlying mass of trachytic conglomerate: it is striped with zones, often of extreme tenuity, of different textures and colours; the harder and darker zones appearing to contain a larger proportion of silica. In another part of the island, there are layers of pearlstone and pitchstone, which in many respects resemble those of Ascension. The zones in the columnar trachyte are generally contorted; they extend un-

---

* MacCulloch states (Classification of Rocks, p. 531), that the exposed surfaces of the pitchstone dikes in Arran are furrowed, "with undulating lines, resembling certain varieties of marbled paper, and which evidently result from some corresponding difference of laminar structure."

† Personal Narrative, vol. i. p. 222.

‡ Geological Transactions, vol. ii. (second series) p. 195.

nterruptedly for a great length in a vertical direction, and apparently parallel to the walls of the dike-like mass. Von Buch* has described at Teneriffe, a stream of lava containing innumerable, thin, plate-like crystals of feldspar, which are arranged like white threads, one behind the other, and which mostly follow the same direction : Dolimieu† also states, that the gray lavas of the modern cone of Vulcanó, which have a vitreous texture, are streaked with parallel white lines : he further describes a solid pumice-stone which possesses a fissile structure, like that of certain micaceous schists. Phonolite, which I may observe is often, if not always, an injected rock, also, often has a fissile structure ; this is generally due to the parallel position of the embedded crystals of feldspar, but sometimes, as at Fernando Noronha, seems to be nearly independent of their presence.‡ From these facts we see, that various rocks of the feldspathic series have either a laminated or fissile structure, and that it occurs both in masses, which have been injected into overlying strata, and in others which have flowed as streams of lava.

The laminæ of the beds, alternating with the obsidian at Ascension, dip at a high angle under the mountain, at the base of which they are situated; and they do not appear as if they had been inclined by violence. A high inclination is common to these beds in Mexico, Peru, and in some of the Italian Islands ;§ on the other hand, in Hungary, the

* Description des Iles Canaries, p. 184.

† Voyage aux Iles de Lipari, pp. 35 and 85.

‡ In this case, and in that of the fissile pumice-stone, the structure is very different from that in the foregoing cases, where the laminæ consist of alternate layers of different composition or texture. In some sedimentary formations, however, which apparently are homogeneous and fissile, as in glassy clay-slate, there is reason to believe, according to D'Aubuisson, that the laminæ are really due to excessively thin alternating, layers of mica.

§ See Phillips' Mineralogy, for the Italian Islands, p. 136. For, Mexico and Peru, see Humboldt's Essai Géognostique. Mr. Edwards, also, describes the high inclination of the obsidian rocks of the Cerro del Navaja in Mexico, in the Proc. of the Geolog. Soc. for June, 1838.

layers are horizontal; the laminæ, also, of some of the lava-streams above referred to, as far as I can understand the descriptions given of them, appear to be highly inclined or vertical. I doubt whether in any of these cases, the laminæ have been tilted into their present position; and in some instances, as in that of the trachyte described by Mr. Scrope, it is almost certain that they have been originally formed with a high inclination. In many of these cases, there is evidence that the mass of liquefied rock has moved in the direction of the laminæ. At Ascension, many of the air-cells have a drawn-out appearance, and are crossed by coarse semi-glassy fibres, in the direction of the laminæ; and some of the layers, separating the sphærulitic globules, have a scored appearance, as if produced by the grating of the globules. I have seen a specimen of zoned obsidian from Mexico, in Mr. Stokes' collection, with the surfaces of the best-defined layers streaked or furrowed with parallel lines; and these lines or streaks precisely resembled those, produced on the surface of a mass of artificial glass by its having been poured out of a vessel. Humboldt, also, has described little cavities, which he compares to the tails of comets, behind sphærulites in laminated obsidian rocks from Mexico, and Mr. Scrope has described other cavities behind fragments embedded in his laminated trachyte, and which he supposes to have been produced during the movement of the mass.* From such facts, most authors have attributed the lamination of these volcanic rocks to their movement whilst liquefied. Although it is easy to perceive, why each

---

* Geological Transactions, vol. ii. (second series) p. 200, &c. These embedded fragments, in some instances, consist of the laminated trachyte broken off and " enveloped in those parts, which still remained liquid." Beudant, also, frequently refers, in his great work on Hungary (tom. iii. p. 386), to trachytic rocks, irregularly spotted with fragments of the same varieties, which in other parts form the parallel ribbons. In these cases, we must suppose, that after part of the molten mass had assumed a laminated structure, a fresh irruption of lava broke up the mass, and involved fragments, and that subsequently the whole became relaminated.

separate air-cell, or each fibre in pumice-stone,* should be
drawn out in the direction of the moving mass; it is by no
means at first obvious why such air-cells and fibres should
be arranged by the movement, in the same planes, in laminæ
absolutely straight and parallel to each other, and often of
extreme tenuity; and still less obvious is it, why such layers
should come to be of slightly different composition and of
different textures.

In endeavouring to make out the cause of the lamination
of these igneous feldspathic rocks, let us return to the facts
so minutely described at Ascension. We there see, that
some of the thinnest layers are chiefly formed by numerous,
exceedingly minute, though perfect, crystals of different
minerals; that other layers are formed by the union of different kinds of concretionary globules, and that the layers
thus formed, often cannot be distinguished from the ordinary
feldspathic and pitchstone layers, composing a large portion
of the entire mass. The fibrous radiating structure of the
sphærulites seems, judging from many analogous cases, to
connect the concretionary and crystalline forces: the separate crystals, also, of feldspar all lie in the same parallel
planes.† These allied forces, therefore, have played an important part in the lamination of the mass, but they cannot be
considered the primary force; for the several kinds of nodules, both the smallest and largest, are internally zoned
with excessively fine shades of colour, parallel to the lamination of the whole; and many of them are, also, externally
marked in the same direction with parallel ridges and furrows, which have not been produced by weathering.

Some of the finest streaks of colour in the stony layers,
alternating with the obsidian, can be distinctly seen to be

---

* Dolimieu's Voyage, p. 64.

† The formation, indeed, of a large crystal of any mineral in a rock
of mixed composition, implies an aggregation of the requisite atoms,
allied to concretionary action. The cause of the crystals of feldspar in
these rocks of Ascension, being all placed lengthways, is probably the
same with that, which elongates and flattens all the brown sphærulitic
globules (which behave like feldspar under the blow-pipe) in this same
direction.

due to an incipient crystallization of the constituent minerals. The extent to which the minerals have crystallized, can, also, be distinctly seen to be connected with the greater or less size, and with the number, of the minute, flattened, crenulated air-cavities or fissures. Numerous facts, as in the case of geodes, and of cavities in silicified wood, in primary rocks, and in veins, show that crystallization is much favoured by space. Hence, I conclude, that, if in a mass of cooling volcanic rock, any cause produced in parallel planes a number of minute fissures or zones of less tension, (which from the pent-up vapours would often be expanded into crenulated air-cavities,) the crystallization of the constituent parts, and probably the formation of concretions, would be superinduced or much favoured in such planes; and thus, a laminated structure of the kind we are here considering would be generated.

That some cause does produce parallel zones of less tension in volcanic rocks, during their consolidation, we must admit in the case of the thin alternate layers of obsidian and pumice described by Humboldt, and of the small, flattened, crenulated air-cells in the laminated rocks of Ascension; for on no other principle can we conceive, why the confined vapours should through their expansion form air-cells or fibres in separate, parallel planes, instead of irregularly throughout the mass. In Mr. Stokes' collection, I have seen a beautiful example of this structure, in a specimen of obsidian from Mexico, which is shaded and zoned, like the finest agate, with numerous, straight parallel layers, more or less opaque and white, or almost perfectly glassy; the degree of opacity and glassiness depending on the number of microscopically minute, flattened air-cells; in this case, it is scarcely possible to doubt but that the mass, to which the fragment belonged, must have been subjected to some, probably prolonged, action, causing the tension slightly to vary in the successive planes.

Several causes appear capable of producing zones of different tension, in masses semi-liquefied by heat. In a fragment of devitrified glass, I have observed layers of sphærulites which appeared, from the manner in which they were abruptly bent, to have been produced by the simple contraction of the mass in the vessel, in which it cooled. In certain dikes on Mount Etna, described by M. Elie de Beaumont,* as bordered by alternating bands of scoriaceous and compact rock, one is led to suppose, that the stretching movement of the surrounding strata, which originally produced the fissures, continued whilst the injected rock remained fluid. Guided, however, by Professor Forbes'† clear description of the zoned structure of glacier-ice, far the most probable explanation of the laminated structure of these feldspathic rocks appears to be, that they have been stretched whilst slowly flowing onwards in a pasty condition,‡ in precisely the same manner as Professor Forbes believes, that the ice of moving glaciers is stretched and fissured. In both cases, the zones may be compared to those in the finest agates; in both, they extend in the direction in which the mass has flowed, and those exposed on the surface are generally vertical: in the ice, the porous laminæ are rendered distinct by the subsequent congelation of infiltrated water, in the stony feldspathic lavas, by subsequent crystalline and concretionary action. The fragment of glassy obsidian in Mr. Stokes' collection, which is zoned with minute air-cells, must strikingly resemble, judging from Professor Forbes' descriptions, a fragment of the zoned ice; and if the rate of cooling and nature of the mass had been favourable to its crystallization or to concretionary action, we should here have had the finest parallel zones of different composition and texture. In glaciers, the lines of

* Mem. pour servir, &c., tom. iv. p. 131.

† Edinburgh New Phil. Journal, 1842, p. 350.

‡ I presume that this is nearly the same explanation which Mr. Scrope had in his mind, when he speaks (Geolog. Transact. vol. ii. second series, p. 228) of the ribboned structure of his trachytic rocks, having arisen, from " a linear extension of the mass, while in a state of imperfect liquidity, coupled with a concretionary process."

porous ice and of minute crevices seem to be due to an incipient stretching, caused by the central parts of the frozen stream moving faster than the sides and bottom, which are retarded by friction: hence, in glaciers of certain forms and towards the lower end of most glaciers, the zones become horizontal. May we venture to suppose that in the feld-spathic lavas with horizontal laminæ, we see an analogous case? All geologists, who have examined trachytic regions, have come to the conclusion, that the lavas of this series have possessed an exceedingly imperfect fluidity; and as it is evident that only matter thus characterized, would be subject to become fissured and to be formed into zones of different tensions, in the manner here supposed, we probably see the reason, why augitic lavas, which appear generally to have possessed a high degree of fluidity, are not,* like the feldspathic lavas, divided into laminæ of different composition and texture. Moreover, in the augitic series, there never appears to be any tendency to concretionary action, which we have seen plays an important part in the lamination of rocks of the trachytic series, or at least in rendering that structure apparent.

Whatever may be thought, of the explanation here advanced of the laminated structure of the rocks of the trachytic series, I venture to call the attention of geologists to the simple fact, that in a body of rock at Ascension, undoubtedly of volcanic origin, layers often of extreme tenuity, quite straight, and parallel to each other, have been produced;—some composed of distinct crystals of quartz and diopside, mingled with amorphous augitic specks and granular feldspar,—others entirely composed of these black augitic specks, with granules of oxide of iron,—and

---

* Basaltic lavas, and many other rocks, are not unfrequently divided into thick laminæ or plates, of the same composition, which are either straight or curved; these being crossed by vertical lines of fissure, sometimes become united into columns. This structure seems related, in its origin, to that by which many rocks, both igneous and sedimentary, become traversed by parallel systems of fissures.

lastly, others formed of crystalline feldspar, in a more or less perfect state of purity, together with numerous crystals of feldspar, placed lengthways. At this island, there is reason to believe, and in some analogous cases, it is certainly known, that the laminæ have originally been formed with their present high inclination. Facts of this nature are manifestly of importance, with relation to the structural origin of that grand series of plutonic rocks, which like the volcanic have undergone the action of heat, and which consist of alternate layers of quartz, feldspar, mica, and other minerals.

CHAPTER IV.

ST. HELENA.

Lavas of the feldspathic, basaltic, and submarine series—Section of Flag-staff Hill and of the Barn—Dikes—Turk's Cap and Prosperous Bays—Basaltic ring—Central crateriform ridge, with an internal ledge and a parapet—Cones of phonolite—Superficial beds of calcareous sandstone—Extinct land-shells—Beds of detritus—Elevation of the land—Denudation—Craters of elevation.

THE whole island is of volcanic origin; its circumference, according to Beatson,* is about twenty-eight miles. The central and largest part consists of rocks of a feldspathic nature, generally decomposed to an extraordinary degree; and when in this state, presenting a singular assemblage of alternating, red, purple, brown, yellow, and white, soft, argillaceous beds. From the shortness of our visit, I did not examine these beds with care; some of them, especially those of the white, yellow, and brown shades, originally existed as streams of lava, but the greater number were probably ejected in the form of scoriæ and ashes: other beds of a purple tint, porphyritic with crystal-shaped patches of a white, soft substance, which are now unctuous, and yield, like wax, a polished streak to the nail, seem once to have existed as solid claystone-porphyryes: the red argillaceous beds generally have a brecciated structure, and no doubt have been formed by the decomposition of scoriæ. Several extensive streams, however, belonging to this series, retain their stony character·

* Governor Beatson's Account of St. Helena.

these are either of a blackish-green colour, with minute
acicular crystals of feldspar, or of a very pale tint, and
almost composed of minute, often scaly, crystals of feldspar,
abounding with microscopical black specks; they are generally compact and laminated; others, however, of similar
composition, are cellular and somewhat decomposed. None
of these rocks contain large crystals of feldspar, or have the
harsh fracture peculiar to trachyte. These feldspathic lavas
and tuffs, are the uppermost or those last erupted; innumerable dikes, however, and great masses of molten rock,
have subsequently been injected into them. They converge,
as they rise, towards the central curved ridge, of which one
point attains the elevation of 2700 feet. This ridge is the
highest land in the island; and it once formed the northern
rim of a great crater, whence the lavas of this series flowed:
from its ruined condition, from the southern half having
been removed, and from the violent dislocation which the
whole island has undergone, its structure is rendered very
obscure.

Basaltic series.—The margin of the island is formed by a
rude circle of great, black, stratified, ramparts of basalt,
dipping seaward, and worn into cliffs, which are often nearly
perpendicular, and vary in height from a few hundred feet to
two thousand. This circle, or rather horse-shoe shaped ring,
is open to the south, and is breached by several other wide
spaces. Its rim or summit generally projects little above
the level of the adjoining inland country; and the more recent
feldspathic lavas, sloping down from the central heights,
generally abut against and overlap its inner margin; on the
north-western side of the island, however, they appear
(judging from a distance) to have flowed over and concealed portions of it. In some parts, where the basaltic
ring has been breached, and the black ramparts stand
detached, the feldspathic lavas have passed between them,
and now overhang the sea-coast in lofty cliffs. The basaltic
rocks are of a black colour and thinly stratified; they are
generally highly vesicular, but occasionally compact; some

of them contain numerous crystals of glassy feldspar and octahedrons of titaniferous iron; others abound with crystals of augite and grains of olivine. The vesicles are frequently lined with minute crystals (of chabasie?) and even become amygdaloidal with them. The streams are separated from each other by cindery matter, or by a bright red, friable, saliferous tuff, which is marked by successive lines like those of aqueous deposition; and sometimes it has an obscure, concretionary structure. The rocks of this basaltic series occur nowhere except near the coast. In most volcanic districts the trachytic lavas are of anterior origin to the basaltic; but here we see, that a great pile of rock, closely related in composition to the trachytic family, has been erupted subsequently to the basaltic strata : the number, however, of dikes, abounding with large crystals of augite, with which the feldspathic lavas have been injected, shows perhaps, some tendency to a return to the more usual order of superposition.

Basal submarine lavas.—The lavas of this basal series lie immediately beneath both the basaltic and feldspathic rocks. According to Mr. Seale,* they may be seen at intervals on the sea-beach round the entire island. In the sections which I examined, their nature varied much; some of the strata abound with crystals of augite; others are of a brown colour, either laminated or in a rubbly condition; and many parts are highly amygdaloidal with calcareous matter. The successive sheets are either closely united together, or are separated from each other by beds of scoriaceous rock and of laminated tuff, frequently containing well rounded fragments. The interstices of these beds are filled with gypsum and salt; the gypsum also, sometimes occurring in thin layers. From the large quantity of these two substances, from the presence of rounded pebbles in the tuffs, and from the abundant amygdaloids, I cannot doubt that

---

* "Geognosy of the Island of St. Helena." Mr. Seale has constructed a gigantic model of St. Helena, well worth visiting, which is now deposited at Addiscombe College, in Surrey.

these basal volcanic strata flowed beneath the sea. This remark ought perhaps to be extended to a part of the super-incumbent basaltic rocks; but on this point, I was not able to obtain clear evidence. The strata of the basal series, wherever I examined them, were intersected by an extra-ordinary number of dikes.

Flagstaff Hill and the Barn.—I will now describe some of the more remarkable sections, and will commence with these two hills, which form the principal external feature on the north-eastern side of the island. The square, angular out-line, and black colour of the Barn, at once show that it belongs to the basaltic series; whilst the smooth, conical figure, and the varied bright tints of Flagstaff Hill, render it equally clear, that it is composed of the softened, feld-spathic rocks. These two lofty hills are connected (as is shown in the accompanying wood-cut) by a sharp ridge,

No. 8.

West.   East.

Flag-staff Hill.   The Barn.
2272 feet high.   2015 feet high.

The double lines represent the basaltic strata; the single, the basal submarine strata; the dotted, the upper feldspathic strata; the dikes are shaded transversely.

which is composed of the rubbly lavas of the basal series. The strata of this ridge dip westward, the inclination be-coming less and less towards the Flagstaff; and the upper feldspathic strata of this hill can be seen, though with some difficulty, to dip conformably to the W.S.W. Close to the Barn, the strata of the ridge are nearly vertical, but are much obscured by innumerable dikes; under this hill, they probably change from being vertical, into being inclined into an opposite direction; for the upper or basaltic strata, which are about 800 or 1000 feet in thickness, are inclined north-eastward, at an angle between thirty and forty degrees.

This ridge, and likewise the Barn and Flagstaff Hills, are

interlaced by dikes, many of which preserve a remarkable parallelism in a N.N.W. and S.S.E. direction. The dikes chiefly consist of a rock, porphyritic with large crystals of augite; others are formed of a fine-grained and brown-coloured trap. Most of these dikes are coated by a glossy layer,* from one to two-tenths of an inch in thickness, which, unlike true pitchstone, fuses into a black enamel; this layer is evidently analogous to the glossy superficial coating of many lava-streams. The dikes can often be followed for great lengths both horizontally and vertically, and they seem to preserve a nearly uniform thickness:† Mr. Seale states, that one near the Barn, in a height of 1260 feet, decreases in width only four inches,—from nine feet at the bottom, to eight feet and eight inches, at the top. On the ridge, the dikes appear to have been guided in their course, to a considerable degree, by the alternating soft and hard strata : they are often firmly united to the harder strata, and they preserve their parallelism for such great lengths, that in very many instances it was impossible to conjecture, which of the beds were dikes, and which streams of lava. The dikes, though so numerous on this ridge, are even more numerous in the valleys a little south of it, and to a degree I never saw equalled any where else : in these valleys they extend in less regular lines, covering the ground with a network, like a spider's web, and with some parts of the surface even appearing to consist wholly of dikes, inter-laced by other dikes.

From the complexity produced by the dikes, from the

* This circumstance has been observed (Lyell, Principles of Geology, vol. iv. chap. x. p. 9) in the dikes of the Atrio del Cavallo, but apparently it is not of very common occurrence. Sir G. Mackenzie, however, states (p. 372, Travels in Iceland) that all the veins in Iceland have a "black vitreous coating on their sides." Capt. Carmichael, speaking of the dikes in Tristan D'Acunha, a volcanic island in the southern Atlantic, says (Linnæan Transactions, vol. xii. p. 485) that their sides, "where they come in contact with the rocks, are invariably in a semi-vitrified state."

† Geognosy of the Island of St. Helena, plate 5.

high inclination and anticlinal dip of the strata of the basal series, which are overlaid, at the opposite ends of the short ridge, by two great masses of different ages and of different composition, I am not surprised that this singular section has been misunderstood. It has even been supposed to form part of a crater; but so far is this from having been the case, that the summit of Flagstaff Hill, once formed the lower extremity of a sheet of lava and ashes, which were erupted from the central, crateriform ridge. Judging from the slope of the contemporaneous streams in an adjoining and undisturbed part of the island, the strata of the Flag-staff Hill, must have been upturned at least twelve hundred feet, and probably much more, for the great truncated dikes on its summit show that it has been largely denuded. The summit of this hill now nearly equals in height the crateri-form ridge; and before having been denuded, it was probably higher than this ridge, from which it is separated by a broad and much lower tract of country: we here, therefore, see that the lower extremity of a set of lava-streams have been tilted up to as great a height as, or perhaps greater height than, the crater, down the flanks of which they originally flowed. I believe that dislocations on so grand a scale are extremely rare* in volcanic districts. The formation of such numbers of dikes in this part of the island, shows that the surface must here have been stretched to a quite extraordinary degree: this stretching, on the ridge between Flagstaff and Barn Hills, probably took place subsequently (though perhaps immediately so) to the strata being tilted; for had the strata at that time extended horizontally, they would in all probability have been fissured and injected transversely, instead of in the planes of their stratification. Although the space between the Barn and Flagstaff Hill, presents a distinct anticlinal line extending north and south, and though most of the dikes range with much regularity in the same

---

* M. Constant Prevost (Mem. de la Soc. Géolog. tom. ii.) observes, that "les produits volcaniques n'ont que localement et rarement même dérangé le sol, à travers lequel ils se sont fait jour."

line, nevertheless, at only a mile due south of the ridge, the strata lie undisturbed. Hence the disturbing force seems to have acted under a point, rather than along a line. The manner in which it has acted, is probably explained by the structure of Little Stony-top, a mountain 2000 feet high, situated a few miles southward of the Barn; we there see, even from a distance, a dark-coloured, sharp, wedge of compact columnar rock, with the bright-coloured feldspathic strata, sloping away on each side from its uncovered apex. This wedge, from which it derives its name of Stony-top, consists of a body of rock, which has been injected whilst liquefied into the overlying strata; and if we may suppose that a similar body of rock lies injected, beneath the ridge connecting the Barn and Flagstaff, the structure there exhibited would be explained.

Turks' Cap and Prosperous Bays.—Prosperous Hill is a great, black, precipitous mountain, situated two miles and a-half south of the Barn, and composed, like it, of basaltic strata. These rest, in one part, on the brown-coloured, porphyritic beds of the basal series, and in another part, on a fissured mass of highly scoriaceous and amygdaloidal rock, which seems to have formed a small point of eruption beneath the sea, contemporaneously with the basal series. Prosperous Hill, like the Barn, is traversed by many dikes, of which the greater number range north and south, and its strata dip, at an angle of about 20°, rather obliquely from the island towards the sea. The space between Prosperous Hill and the Barn, as represented in this wood-cut, consists

No. 9.

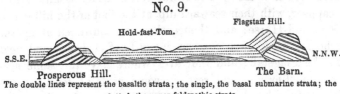

The double lines represent the basaltic strata; the single, the basal submarine strata; the dotted, the upper feldspathic strata.

of lofty cliffs, composed of the lavas of the upper or feldspathic series, which rest, though unconformably, on the

basal submarine strata, as we have seen that they do at
Flagstaff Hill. Differently, however, from in that hill, these
upper strata are nearly horizontal, gently rising towards the
interior of the island ; and they are composed of greenish-
black, or more commonly, pale-brown, compact lavas,
instead of softened and highly coloured matter. These
brown-coloured, compact lavas, consist almost entirely of
small glimmering scales, or of minute acicular crystals,
of feldspar, placed close by the side of each other, and
abounding with minute black specks, apparently of hornblende. The basaltic strata of Prosperous Hill project only
a little above the level of the gently-sloping, feldspathic
streams, which wind round and abut against their upturned
edges. The inclination of the basaltic strata seems to be too
great, to have been caused by their having flowed down a
slope, and they must have been tilted into their present position, before the eruption of the feldspathic streams.

Basaltic ring.—Proceeding round the island, the lavas of
the upper series, southward of Prosperous Hill, overhang
the sea in lofty precipices. Further on, the headland, called
Great Stony-top, is composed, as I believe, of basalt; as is
Long Range Point, on the inland side of which, the coloured
beds abut. On the southern side of the island, we see the
basaltic strata of the South Barn, dipping obliquely seaward
at a considerable angle; this headland, also, stands a little
above the level of the more modern, feldspathic lavas. Further on, a large space of coast, on each side of Sandy Bay,
has been much denuded, and there seems to be left only the
basal wreck of the great, central crater. The basaltic strata
reappear, with their seaward dip, at the foot of the hill called
Man-and-Horse; and thence they are continued along the
whole north-western coast to Sugar-Loaf Hill, situated near
to the Flagstaff; and they everywhere have the same seaward inclination, and rest, in some parts at least, on the
lavas of the basal series. We thus see that the circumference of the island, is formed by a much-broken ring, or
rather a horse-shoe of basalt, open to the south, and inter-

rupted on the eastern side by many wide breaches. The breadth of this marginal fringe on the north-western side, where alone it is at all perfect, appears to vary from a mile to a mile and a-half. The basaltic strata, as well as those of the subjacent basal series, dip, with a moderate inclination, where they have not been subsequently disturbed, towards the sea. The more broken state of the basaltic ring round the eastern half, compared with the western half of the island, is evidently due to the much greater denuding power of the waves on the eastern or windward side, as is shown by the greater height of the cliffs on that side, than to lee-ward. Whether the margin of basalt was breached, before or after the eruption of the lavas of the upper series, is doubtful; but as separate portions of the basaltic ring appear to have been tilted before that event, and from other reasons, it is more probable, that some at least of the breaches were first formed. Reconstructing in imagination, as far as is possible, the ring of basalt, the internal space or hollow, which has since been filled up with the matter erupted from the great central crater, appears to have been of an oval figure, eight or nine miles in length by about four miles in breadth, and with its axis directed in a N.E. and S.W. line, coincident with the present longest axis of the island.

The central curved ridge.—This ridge consists, as before remarked, of gray feldspathic lavas, and of red, brecciated, argillaceous tuffs, like the beds of the upper coloured series. The gray lavas contain numerous, minute, black, easily fusible specks; and but very few large crystals of feldspar. They are generally much softened; with the exception of this character, and of being in many parts highly cellular, they are quite similar to those great sheets of lava which overhang the coast at Prosperous Bay. Considerable intervals of time appear to have elapsed, judging from the marks of denudation, between the formation of the successive beds, of which this ridge is composed. On the steep northern slope, I observed in several sections a much worn undulating surface of red tuff, covered by gray, decomposed, feldspathic lavas,

G

with only a thin earthy layer interposed between them.   In
an adjoining part, I noticed a trap-dike, four feet wide, cut
off and covered up by the feldspathic lava, as is represented
in the wood-cut.   The ridge ends on the eastern side in

No. 10.

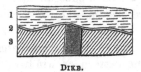

DIKE.

1—Gray feldspathic lava.
2—A layer, one inch in thickness, of a reddish earthy matter.
3—Brecciated, red, argillaceous tuff.

a hook, which is not represented clearly enough in any map
which I have seen; towards the western end, it gradually
slopes down and divides into several subordinate ridges.
The best defined portion between Diana's Peak and
Nest Lodge, which supports the highest pinnacles in the
island varying from 2000 to 2700 feet, is rather less than
three miles long in a straight line.   Throughout this space
the ridge has a uniform appearance and structure; its curvature resembles that of the coast-line of a great bay, being
made up of many smaller curves, all open to the south.
The northern and outer side is supported by narrow ridges
or buttresses, which slope down to the adjoining country.
The inside is much steeper, and is almost precipitous; it is
formed of the basset edges of the strata, which gently de-
cline outwards.   Along some parts of the inner side, a little
way beneath the summit, a flat ledge extends, which imitates
in outline the smaller curvatures of the crest.   Ledges of
this kind occur not unfrequently within volcanic craters, and
their formation seems to be due to the sinking down of a
level sheet of hardened lava, the edges of which remain
(like the ice round a pool, from which the water has been
drained) adhering to the sides.*

* A most remarkable instance of this structure is described in Ellis'
Polynesian Researches (second edit.), where an admirable drawing is

In some parts, the ridge is surmounted by a wall or parapet, perpendicular on both sides. Near Diana's Peak this wall is extremely narrow. At the Galapagos Archipelago I observed parapets, having a quite similar structure and appearance, surmounting several of the craters; one, which I more particularly examined, was composed of glossy, red scoriæ firmly cemented together; being externally perpendicular, and extending round nearly the whole circumference of the crater, it rendered it almost inaccessible. The Peak of Teneriffe and Cotopaxi, according to Humboldt, are similarly constructed; he states* that "at their summits a circular wall surrounds the crater, which wall, at a distance, has the appearance of a small cylinder placed on a truncated cone. On Cotopaxi† this peculiar structure is visible to the naked eye at more than 2000 toises' distance; and no person has ever reached its crater. On the Peak of Teneriffe, the parapet is so high, that it would be impossible to reach the *caldera*, if on the eastern side there did not exist a breach." The origin of these circular parapets, is probably due to the heat or vapours from the crater, penetrating and hardening the sides to a nearly equal depth, and afterwards to the mountain being slowly acted on by the weather, which would leave the hardened part, projecting in the form of a cylinder or circular parapet.

From the points of structure in the central ridge, now enumerated,—namely, from the convergence towards it of the beds of the upper series,—from the lavas there becoming highly cellular,—from the flat ledge, extending along its inner and precipitous side, like that within some still active craters,—from the parapet-like wall on its summit,—and lastly, from its peculiar curvature, unlike that of any common line of elevation, I cannot doubt that this curved ridge forms the last remnant of a great crater. In endeavouring,

given of the successive ledges or terraces, on the borders of the immense crater at Hawaii, in the Sandwich Islands.

* Personal Narrative, vol. i. p. 171.
† Humboldt's Picturesque Atlas, folio, pl. 10.

however, to trace its former outline, one is soon baffled; its western extremity gradually slopes down, and branching into other ridges, extends to the sea-coast; the eastern end is more curved, but it is only a little better defined. Some appearances lead me to suppose, that the southern wall of the crater joined the present ridge near Nest Lodge; in this case the crater must have been nearly three miles long, and about a mile and a-half in breadth. Had the denudation of the ridge, and the decomposition of its constituent rocks, proceeded a few steps further, and had this ridge, like several other parts of the island, been broken up by great dikes and masses of injected matter, we should in vain have endeavoured to discover its true nature. Even now we have seen, that at Flagstaff Hill, the lower extremity and most distant portion of one sheet of the erupted matter, has been upheaved, to as great a height as the crater down which it flowed, and probably even to a greater height. It is interesting thus to trace the steps, by which the structure of a volcanic district becomes obscured, and finally obliterated: so near to this last stage is St. Helena, that I believe no one has hitherto suspected, that the central ridge or axis of the island, is the last wreck of the crater, whence the most modern volcanic streams were poured forth.

The great hollow space or valley southward of the central curved ridge, across which the half of the crater must once have extended, is formed of bare, water-worn hillocks and ridges of red, yellow, and brown rocks, mingled together in chaos-like confusion, interlaced by dikes, and without any regular stratification. The chief part consists of red decomposing scoriæ, associated with various kinds of tuff and yellow argillaceous beds, full of broken crystals, those of augite being particularly large. Here and there masses of highly cellular and amygdaloidal lavas protrude. From one of the ridges, in the midst of the valley, a conical precipitous hill, called Lot, boldly stands up, and forms a most singular and conspicuous object. It is composed of phonolite, divided in one part into great curved laminæ, in another, into angular concretionary balls, and in a third part, into outwardly radiating columns. At its base the strata of lava, tuff, and scoriæ, dip away on all sides :* the uncovered portion is 197† feet in height, and its horizontal section gives an oval figure. The phonolite is of a greenish-gray colour, and is full of minute acicular crystals of feldspar ; in most parts it has a conchoidal fracture, and is sonorous, yet it is crenulated with minute air-cavities. In a S.W. direction from Lot, there are some other remarkable columnar pinnacles, but of a less regular shape, namely, Lot's Wife, and the Asses' Ears, composed of allied kinds of rock. From their flattened shape, and their relative position to each other, they are evidently connected on the same line of fissure. It is, moreover, remarkable, that this same N.E. and S.W. line, joining Lot and Lot's Wife, if prolonged, would intersect Flagstaff Hill, which, as before stated, is crossed by numerous dikes running in this direction, and which has a disturbed structure, rendering it probable that a great body of once fluid rock lies injected beneath it.

In this same great valley, there are several other conical masses of injected rock, (one, I observed, was composed of compact greenstone) some of which are not connected, as far as is apparent, with any line of dike; whilst others are obviously thus connected. Of these dikes, three or four great lines stretch across the valley in a N.E and S.W. direction, parallel to that one connecting the Asses' Ears, Lot's Wife, and probably Lot. The number of these masses of injected rock, is a remarkable feature in the geology of St. Helena. Beside those just mentioned, and the hypothetical one beneath Flagstaff Hill, there is Little Stony-top

* Abich, in his Views of Vesuvius (plate vi.), has shown the manner in which beds, under nearly similar circumstances, are tilted up. The upper beds are more turned up than the lower ; and he accounts for this, by showing that the lava insinuates itself horizontally between the lower beds.

† This height is given by Mr. Seale, in his Geognosy of the island. The height of the summit above the level of the sea, is said to be 1444 feet.

and others, as I have reason to believe, at the Man-and-Horse, and at High Hill. Most of these masses, if not all of them, have been injected, subsequently to the last volcanic eruptions from the central crater. The formation of conical bosses of rock on lines of fissure, the walls of which are in most cases parallel, may probably be attributed to inequalities in the tension, causing small transverse fissures; and at these points of intersection, the edges of the strata would naturally yield, and be easily turned upwards. Finally, I may remark, that hills of phonolite everywhere are apt[*] to assume singular and even grotesque shapes, like that of Lot: the peak at Fernando Noronha offers an instance; at St. Jago, however, the cones of phonolite, though tapering, have a regular form. Supposing, as seems probable, that all such hillocks or obelisks have originally been injected, whilst liquefied, into a mould formed by yielding strata, as certainly has been the case with Lot, how are we to account for the frequent abruptness and singularity of their outlines, compared with similarly injected masses of greenstone and basalt? Can it be due to a less perfect degree of fluidity, which is generally supposed to be characteristic of the allied, trachytic lavas?

Superficial deposits.—Soft calcareous sandstone occurs in extensive, though thin, superficial beds, both on the northern and southern shores of the island. It consists of very minute, equal-sized, rounded particles of shells, and other organic bodies, which partially retain their yellow, brown, and pink colours, and occasionally, though very rarely, present an obscure trace of their original external forms. I in vain endeavoured to find a single unrolled fragment of a shell. The colour of the particles, is the most obvious character, by which their origin can be recognized, the tints being affected (and an odour produced) by a moderate heat, in the same manner as in fresh shells. The particles are cemented together, and are mingled with some earthy matter: the

* D'Aubuisson, in his Traité de Géognosie (tom. ii. p. 540), particularly remarks that this is the case.

purest masses, according to Beatson, contain 70 per cent. of carbonate of lime. The beds, varying in thickness from two or three feet to fifteen feet, coat the surface of the ground; they generally lie on that side of the valley which is protected from the wind, and they occur at the height of several hundred feet above the level of the sea. Their position is the same, which sand, if now drifted by the trade-wind, would occupy; and no doubt they thus originated, which explains the equal size and minuteness of the particles, and likewise the entire absence of whole shells, or even of moderately-sized fragments. It is remarkable that at the present day, there are no shelly beaches on any part of the coast, whence calcareous dust could be drifted and winnowed; we must, therefore, look back to a former period, when, before the land was worn into the present great precipices, a shelving coast, like that of Ascension, was favourable to the accumulation of shelly detritus. Some of the beds of this limestone are between 600 and 700 feet above the sea; but part of this height may possibly be due to an elevation of the land, subsequent to the accumulation of the calcareous sand.

The percolation of rain-water has consolidated parts of these beds into a solid rock, and has formed masses of dark brown, stalagmitic limestone. At the Sugar-Loaf quarry, fragments of rock on the adjoining slopes,* have been thickly coated by successive fine layers of calcareous matter. It is singular, that many of these pebbles have their entire surfaces coated, without any point of contact having been left uncovered; hence, these pebbles must have been lifted up by the slow deposition between them, of the successive films of carbonate of lime. Masses of white, finely oolitic rock are attached to the outside of some of these coated

---

* In the earthy detritus on several parts of this hill, irregular masses of very impure, crystallized sulphate of lime occur. As this substance is now being abundantly deposited by the surf at Ascension, it is possible that these masses may thus have originated; but if so, it must have been at a period, when the land stood at a much lower level. This earthy selenite is now found at a height of between 600 and 700 feet.

pebbles. Von Buch has described a compact limestone at
Lanzarote, which seems perfectly to resemble the stalagmitic
deposition just mentioned : it coats pebbles, and in parts is
finely oolitic : it forms a far-extended layer, from one inch to
two or three feet in thickness, and it occurs at the height of
800 feet above the sea, but only on that side of the island
exposed to the violent north-western winds. Von Buch
remarks,* that it is not found in hollows, but only on the
unbroken and inclined surfaces of the mountain. He believes, that it has been deposited by the spray which is
borne over the whole island by these violent winds. It
appears, however, to me much more probable that it has
been formed, as at St. Helena, by the percolation of water
through finely comminuted shells : for when sand is blown
on a much exposed coast, it always tends to accumulate on
broad, even surfaces, which offer a uniform resistance to the
winds. At the neighbouring island, moreover, of Feurte-
ventura,† there is an earthy limestone, which, according to
Von Buch, is quite similar to specimens which he has seen
from St. Helena, and which he believes to have been formed
by the drifting of shelly detritus.

The upper beds of the limestone, at the above-mentioned
quarry on the Sugar-Loaf Hill, are softer, finer-grained and
less pure, than the lower beds. They abound with fragments of land-shells, and with some perfect ones ; they contain, also, the bones of birds, and the large eggs,‡ apparently
of water-fowl. It is probable that these upper beds remained long in an unconsolidated form, during which time,
these terrestrial productions were embedded. Mr. G. R.
Sowerby has kindly examined three species of land-shells,
which I procured from this bed, and his descriptions are

* Description des Isles Canàries, p. 293.
† Idem, pp. 314 and 374.
‡ Colonel Wilkes, in a catalogue presented with some specimens to
the Geological Society, states that as many as ten eggs were found by
one person. Dr. Buckland has remarked (Geolog. Trans. vol. v. p. 474)
on these eggs.

given in the Appendix. One of them is a Succinea, identical
with a species, now living abundantly on the island : the two
others, namely, *Cochlogena fossilis,* and *Helix biplicata,* are
not known in a recent state: the latter species was also
found in another and different locality, associated with a
species of Cochlogena, which is undoubtedly extinct.

Beds of extinct land-shells.—Land-shells, all of which appear to be species now extinct, occur embedded in earth, in
several parts of the island. The greater number have been
found at a considerable height on Flagstaff Hill. On the
N.W. side of this hill, a rain-channel exposes a section of
about twenty feet in thickness, of which the upper part
consists of black vegetable mould, evidently washed down
from the heights above, and the lower part of less black
earth, abounding with young and old shells, and with their
fragments: part of this earth is slightly consolidated by
calcareous matter, apparently due to the partial decomposition of some of the shells. Mr. Seale, an intelligent resi-
dent, who first called attention to these shells, gave me a
large collection from another locality, where the shells
appear to have been embedded in very black earth. Mr. G.
R. Sowerby has examined these shells, and has described
them in the Appendix. There are seven species, namely,
one Cochlogena, two species of the genus Cochlicopa, and
four of Helix: none of these are known in a recent state, or
have been found in any other country. The smaller species
were picked out of the inside of the large shells of the
Cochlogena auris-vulpina. This last-mentioned species is in
many respects a very singular one; it was classed, even by
Lamarck, in a marine genus, and having thus been mistaken
for a sea-shell, and the smaller accompanying species having
been overlooked, the exact localities where it was found,
have been measured, and the elevation of this island thus
deduced! It is very remarkable that all the shells of this
species found by me in one spot, form a distinct variety, as
described by Mr. Sowerby, from those procured from another locality by Mr. Seale. As this Cochlogena is a large

and conspicuous shell, I particularly enquired from several intelligent countrymen whether they had ever seen it alive; they all assured me that they had not, and they would not even believe that it was a land animal: Mr. Seale, moreover, who was a collector of shells all his life at St. Helena, never met with it alive. Possibly some of the smaller species may turn out to be yet living kinds; but, on the other hand, the two land-shells which are now living on the island in great numbers, do not occur embedded, as far as is yet known, with the extinct species. I have shown in my Journal,* that the extinction of these land-shells possibly may not be an ancient event; as a great change took place in the state of the island about 120 years ago, when the old trees died, and were not replaced by young ones, these being destroyed by the goats and hogs, which had run wild in numbers, from the year 1502. Mr. Seale states, that on Flagstaff Hill, where we have seen that the embedded land-shells are especially numerous, traces are everywhere discoverable, which plainly indicate that it was once thickly clothed with trees; at present not even a bush grows there. The thick bed of black vegetable mould which covers the shell-bed, on the flanks of this hill, was probably washed down from the upper part, as soon as the trees perished, and the shelter afforded by them was lost.

Elevation of the land.—Seeing that the lavas of the basal series, which are of submarine origin, are raised above the level of the sea, and at some places to the height of many hundred feet, I looked out for superficial signs of the elevation of the land. The bottoms of some of the gorges, which descend to the coast, are filled up to the depth of about a hundred feet, by rudely divided layers of sand, muddy clay, and fragmentary masses; in these beds, Mr. Seale has found the bones of the tropic-bird and of the albatross; the former now rarely, and the latter never visiting the island. From the difference between these layers, and the sloping piles of detritus which rest on them, I suspect that they were

* Journal of Researches, p. 582.

deposited, when the gorges stood beneath the sea. Mr. Seale, moreover, has shown that some of the fissure-like gorges,* become, with a concave outline, gradually rather wider at the bottom, than at the top; and this peculiar structure was probably caused by the wearing action of the sea, when it entered the lower part of these gorges. At greater heights, the evidence of the rise of the land is even less clear: nevertheless, in a bay-like depression on the table-land behind Prosperous Bay, at the height of about 1000 feet, there are flat-topped masses of rock, which it is scarcely conceivable, could have been insulated from the surrounding and similar strata, by any other agency than the denuding action of a sea-beach. Much denudation, indeed, has been effected at great elevations, which it would not be easy to explain by any other means: thus, the flat summit of the Barn, which is 2000 feet high, presents, according to Mr. Seale, a perfect net-work of truncated dikes; on hills like the Flagstaff, formed of soft rock, we might suppose that the dikes had been worn down and cut off by meteoric agency, but we can hardly suppose this possible with the hard, basaltic strata of the Barn.

Coast denudation.—The enormous cliffs, in many parts between 1000 and 2000 feet in height, with which this prison-like island is surrounded, with the exception of only a few places, where narrow valleys descend to the coast, is the most striking feature in its scenery. We have seen that portions of the basaltic ring, two or three miles in length by one or two miles in breadth, and from one to two thousand feet in height, have been wholly removed. There are, also, ledges and banks of rock, rising out of profoundly deep water, and distant from the present coast between three and four miles, which, according to Mr. Seale, can be traced to the shore, and are found to be the continuations of certain well-known great dikes. The swell of the Atlantic ocean has obviously been the active power in forming these cliffs;

---

* A fissure-like gorge, near Stony-top, is said by Mr. Seale to be 840 feet deep, and only 115 feet in width.

and it is interesting to observe that the lesser, though still great, height of the cliffs on the leeward and partially protected side of the island, (extending from the Sugar-Loaf Hill to South West Point,) corresponds with the lesser degree of exposure. When reflecting on the comparatively low coasts of many volcanic islands, which also stand exposed in the open ocean, and are apparently of considerable antiquity, the mind recoils from an attempt to grasp the number of centuries of exposure, necessary to have ground into mud and to have dispersed, the enormous cubic mass of hard rock, which has been pared off the circumference of this island. The contrast in the superficial state of St. Helena, compared with the nearest island, namely, Ascension, is very striking. At Ascension, the surface of the lava-streams are glossy, as if just poured forth, their boundaries are well defined, and they can often be traced to perfect craters, whence they were erupted; in the course of many long walks, I did not observe a single dike; and the coast round nearly the entire circumference, is low, and has been eaten back (though too much stress must not be placed on this fact, as the island may have been subsiding) into a little wall only from ten to thirty feet high. Yet during the 340 years, since Ascension has been known, not even the feeblest signs of volcanic action have been recorded.* On the other hand, at St. Helena, the course of no one stream of lava can be traced, either by the state of its boundaries or of its superficies ; the mere wreck of one great crater is left ; not the valleys only, but the surface of some of the

---

* In the Nautical Magazine for 1835, p. 642, and for 1838, p. 361, and in the Comptes Rendus, April, 1838, accounts are given of a series of volcanic phenomena—earthquakes—troubled water—floating scoriæ and columns of smoke—which have been observed at intervals since the middle of the last century, in a space of open sea between longitudes 20° and 22° west, about half a degree south of the equator. These facts seem to show, that an island or an archipelago is in process of formation in the middle of the Atlantic: a line joining St. Helena and Ascension, prolonged, intersects this slowly nascent focus of volcanic action.

highest hills, are interlaced by worn-down dikes, and, in many places, the denuded summits of great cones of injected rock stand exposed and naked ; lastly, as we have seen, the entire circuit of the island has been deeply worn back into the grandest precipices.

Craters of Elevation.

There is much resemblance in structure and in geological history between St. Helena, St. Jago, and Mauritius. All three islands are bounded (at least in the parts, which I was able to examine) by a ring of basaltic mountains, now much broken, but evidently once continuous. These mountains have, or apparently once had, their escarpements steep towards the interior of the island, and their strata dip outwards. I was able to ascertain, only in a few cases, the inclination of the beds ; nor was this easy, for the stratification was generally obscure, except when viewed from a distance. I feel, however, little doubt that according to the researches of M. Elie de Beaumont, their average inclination is greater than that, which they could have acquired, considering their thickness and compactness, by flowing down a sloping surface. At St. Helena, and at St. Jago, the basaltic strata rest on older and probably submarine beds, of different composition. At all three islands, deluges of more recent lavas have flowed from the centre of the island, towards and between the basaltic mountains ; and at St. Helena, the central platform has been filled up by them. All three islands have been raised in mass. At Mauritius, the sea, within a late geological period, must have reached to the foot of the basaltic mountains, as it now does at St. Helena ; and at St. Jago, it is cutting back the intermediate plain towards them. In these three islands, but especially at St. Jago and at Mauritius, when standing on the summit of one of the old basaltic mountains, one looks in vain towards the centre of the island,—the point, towards which the strata beneath one's feet and of the mountains on each side, rudely converge,—for a source whence these strata could have been erupted; but one sees only a vast hollow platform stretched beneath, or piles of matter of more recent origin.

These basaltic mountains come, I presume, into the class of Craters of elevation : it is immaterial whether the rings were ever completely formed, for the portions which now exist, have so uniform a structure, that, if they do not form fragments of true craters, they cannot be classed with ordinary lines of elevation. With respect to their origin, after having read the works of Mr. Lyell,* and of MM. C. Prevost and Virlet, I cannot believe, that the great central hollows have been formed by a simple dome-shaped elevation, and the consequent arching of the strata. On the other hand, I have very great difficulty in admitting, that these basaltic mountains are merely the basal fragments of great volcanos, of which the summits have either been blown off, or more probably swallowed up by subsidence. These rings are in some instances so immense, as at St. Jago and at Mauritius, and their occurrence is so frequent, that I can hardly persuade myself to adopt this explanation. Moreover, I suspect that the following circumstances, from their frequent concurrence, are someway connected together,—a connection not implied in either of the above views ; namely, first, the broken state of the ring, showing that the now detached portions have been exposed to great denudation, and in some cases perhaps, rendering it probable that the ring never was entire; secondly, the great amount of matter erupted from the central area, after or during, the formation of the ring ; and thirdly, the elevation of the district in mass. As far as relates to the inclination of the strata being greater than that, which the basal fragments of ordinary volcanos would naturally possess, I can readily believe that this inclination might have been slowly acquired by that amount of elevation, of which, according to M. Elie de Beaumont, the numerous upfilled fissures or dikes are the evidence and the measure,— a view equally novel and important, which we owe to the researches of that geologist on Mount Etna.

* Principles of Geology (fifth edit.), vol. ii. p. 171.

A conjecture, including the above circumstances, occurred to me, when,—with my mind fully convinced from the phenomena of 1835 in South America,* that the forces, which eject matter from volcanic orifices and raise continents in mass, are identical,—I viewed that part of the coast of St. Jago, where the horizontally upraised, calcareous stratum dips into the sea, directly beneath a cone of subsequently erupted lava. The conjecture is, that during the slow elevation of a volcanic district or island, in the centre of which one or more orifices continue open, and thus relieve the subterranean forces, the borders are elevated more than the central area; and that the portions thus upraised, do not slope gently into the central, less elevated area, as does the calcareous stratum under the cone at St. Jago, and as does a large part of the circumference of Iceland,† but that they

* I have given a detailed account of these phenomena, in a paper read before the Geological Society in March, 1838. At the instant of time, when an immense area was convulsed and a large tract elevated, the districts immediately surrounding several of the great vents in the Cordillera remained quiescent; the subterranean forces being apparently relieved by the eruptions, which then recommenced with great violence. An event of somewhat the same kind, but on an infinitely smaller scale, appears to have taken place, according to Abich (Views of Vesuvius, plates i. and ix.), within the great crater of Vesuvius, where a platform on one side of a fissure was raised in mass twenty feet, whilst on the other side, a train of small volcanos burst forth in eruption.

† It appears, from information communicated to me in the most obliging manner by M. E. Robert, that the circumferential parts of Iceland, which are composed of ancient basaltic strata alternating with tuff, dip inland, thus forming a gigantic saucer. M. Robert found that this was the case, with a few and quite local exceptions, for a space of coast several hundred miles in length. I find this statement corroborated, as far as regards one place, by Mackenzie, in his Travels (p. 377), and in another place by some MS. notes kindly lent me by Dr. Holland. The coast is deeply indented by creeks, at the head of which the land is generally low. M. Robert informs me, that the inwardly dipping strata appear to extend as far as this line, and that their inclination usually corresponds with the slope of the surface, from the high coast-mountains to the low land at the head of these creeks. In the section described by Sir G. Mackenzie, the dip is 12°. The

are separated from it by curved faults. We might expect
from what we see along ordinary faults, that the strata
on the upraised side, already dipping outwards from their
original formation as lava-streams, would be tilted from the
line of fault, and thus have their inclination increased.
According to this hypothesis, which I am tempted to extend
only to some few cases, it is not probable that the ring
would ever be formed quite perfect; and from the elevation
being slow, the upraised portions would generally be exposed to much denudation, and hence the ring become
broken; we might also expect to find occasional inequalities
in the dip of the upraised masses, as is the case at St. Jago.
By this hypothesis, the elevation of the districts in mass, and
the flowing of deluges of lava from the central platforms,
are likewise connected together.   On this view, the marginal
basaltic mountains of the three foregoing islands, might still
be considered as forming, " Craters of elevation;" the kind
of elevation implied having been slow, and the central
hollow or platform having been formed, not by the arching
of the surface, but simply by that part having been upraised
to a less height.

interior parts of the island chiefly consist, as far as is known, of
recently erupted matter. The great size, however, of Iceland, equalling
the bulkiest part of England, ought perhaps to exclude it from the
class of islands we have been considering; but I cannot avoid suspect-
ing that if the coast-mountains, instead of gently sloping into the less
elevated central area, had been separated from it by irregularly curved
faults, the strata would have been tilted seaward, and a " crater of
elevation," like that of St. Jago or that of Mauritius, but of much vaster
dimensions, would have been formed.   I will only further remark, that
the frequent occurrence of extensive lakes at the foot of large volcanos,
and the frequent association of volcanic and fresh-water strata, seem to
indicate that the areas around volcanos are apt to be depressed beneath
the general level of the adjoining country, either from having been
less elevated, or from the effects of subsidence.

CHAPTER V.

GALAPAGOS ARCHIPELAGO.

Chatham Island— Craters composed of a peculiar kind of ˙tuff—Small basaltic craters, with hollows at their bases—Albemarle Island, fluid lavas, their composition—Craters of tuff, inclination of their exterior diverging strata, and structure of their interior converging strata— James Island, segment of a small basaltic crater ; fluidity and composition of its lava streams, and of its ejected fragments—Concluding remarks on the craters of tuff, and on the breached condition of their southern sides—Mineralogical composition of the rocks of the archipelago—Elevation of the land—Direction of the fissures of eruption.

THIS archipelago is situated under the Equator, at a distance of between five and six hundred miles from the west coast of South America. It consists of five principal islands, and of several small ones, which together are equal in area,* but not in extent of land, to Sicily conjointly with the Ionian islands. They are all volcanic: on two, craters have been seen in eruption, and on several of the other islands, streams of lava have a recent appearance. The larger islands are chiefly composed of solid rock, and they rise with a tame outline, to a height of between one and four thousand feet. They are sometimes, but not generally, surmounted by one principal orifice. The craters vary in size from mere spiracles to huge caldrons, several miles in circumference; they are extraordinarily numerous, so that I should think, if enumerated, they would be found to exceed two thousand ;

* I exclude from this measurement, the small volcanic islands of Culpepper and Wenman, lying seventy miles northward of the group. Craters were visible on all the islands of the group, except on Towers Island, which is one of the lowest; this island is, however formed of volcanic rocks.

H

they are formed either of scoriæ and lava, or of a brown
coloured tuff; and these latter craters are in several respects
remarkable. The whole group was surveyed by the officers
of the Beagle. I visited myself four of the principal islands,
and received specimens from all the others. Under the head
of the different islands, I will describe only that which
appears to me deserving of attention.

No. 11.

GALAPAGOS ARCHIPELAGO.

CHATHAM ISLAND. *Craters composed of a singular kind of
tuff.*—Towards the eastern end of this island, there occur two
craters, composed of two kinds of tuff; one kind being friable,
like slightly consolidated ashes; and the other compact, and
of a different nature from any thing, which I have met with
described. This latter substance, where it is best characterized,
is of a yellowish-brown colour, translucent, and with a lustre
somewhat resembling resin; it is brittle, with an angular,
rough, and very irregular fracture, sometimes, however, being
slightly granular, and even obscurely crystalline: it can

readily be scratched with a knife, yet some points are hard enough just to mark common glass; it fuses with ease into a blackish-green glass. The mass contains numerous broken crystals of olivine and augite, and small particles of black and brown scoriæ: it is often traversed by thin seams of calcareous matter. It generally affects a nodular or concretionary structure. In a hand specimen, this substance would certainly be mistaken for a pale and peculiar variety of pitchstone; but when seen in mass, its stratification, and the numerous layers of fragments of basalt, both angular and rounded, at once render its subaqueous origin evident. An examination of a series of specimens, shows that this resin-like substance, results from a chemical change on small particles of pale and dark-coloured, scoriaceous rocks; and this change could be distinctly traced in different stages, round the edges of even the same particle. The position near the coast, of all the craters composed of this kind of tuff or peperino, and their breached condition, renders it probable that they were all formed, when standing immersed in the sea; considering this circumstance, together with the remarkable absence of large beds of ashes in the whole archipelago, I think it highly probable, that much the greater part of the tuff has originated, from the trituration of fragments of the gray, basaltic lavas, in the mouths of craters standing in the sea. It may be asked, whether the heated water within these craters, has produced this singular change in the small scoriaceous particles, and given to them their translucent, resin-like fracture? Or has the associated lime played any part in this change? I ask these questions, from having found at St. Jago, in the Cape de Verde Islands, that where a great stream of molten lava has flowed over a calcareous bottom, into the sea, the outermost film, which in other parts resembles pitchstone, is changed, apparently by its contact with the carbonate of lime, into a resin-like substance, precisely like the best characterized specimens of the tuff from this archipelago.*

* The concretions containing lime, which I have described at

H 2

To return to the two craters: one of them stands at the
distance of a league from the coast, the intervening tract
consisting of a calcareous tuff, apparently of submarine
origin. This crater consists of a circle of hills, some of
which stand quite detached, but all have a very regular, quâ-
quâ versal dip, at an inclination of between thirty and forty
degrees. The lower beds, to the thickness of several hundred
feet, consist of the resin-like stone, with embedded fragments
of lava. The upper beds, which are between thirty and
forty feet in thickness, are composed of a thinly stratified,
fine-grained, harsh, friable, brown-coloured tuff, or peperino.* A central mass without any stratification, which
must formerly have occupied the hollow of the crater, but is
now attached only to a few of the circumferential hills,
consists of a tuff, intermediate in character between that
with a resin-like, and that with an earthy fracture. This
mass contains white calcareous matter in small patches.
The second crater (520 feet in height) must have existed,
until the eruption of a recent, great stream of lava, as a
separate islet; a fine section, worn by the sea, shows a grand
funnel-shaped mass of basalt, surrounded by steep, sloping,
flanks of tuff, having in parts an earthy, and in others, a
semi-resinous fracture.   The tuff is traversed by several
broad, vertical dikes, with smooth and parallel sides, which
I did not doubt were formed of basalt, until I actually broke
off fragments. These dikes, however, consist of tuff like
that of the surrounding strata, but more compact, and with
a smoother fracture; hence we must conclude, that fissures
were formed and filled up with the finer mud or tuff from

Ascension, as formed in a bed of ashes, present some degree of resemblance to this substance, but they have not a resinous fracture.   At St.
Helena, also, I found veins of a somewhat similar, compact, but non-
resinous substance, occurring in a bed of pumiceous ashes, apparently
free from calcareous matter : in neither of these cases could heat have
acted.
    * Those geologists who restrict the term of tuff, to ashes of a white
colour, resulting from the attrition of feldspathic lavas, would call these
brown-coloured strata "peperino."

the crater, before its interior was occupied, as it now is, by a solidified pool of basalt. Other fissures have been subsequently formed, parallel to these singular dikes, and are merely filled with loose rubbish. The change from ordinary scoriaceous particles to the substance with a semi-resinous fracture, could be clearly followed in portions of the compact tuff of these dikes.

At the distance of a few miles from these two craters, stands the Kicker rock, or islet, remarkable from its singular form. It is unstratified, and is composed of compact tuff, in parts having the resin-like fracture. It is probable that

No. 12.

THE KICKER ROCK,—400 feet high.

this amorphous mass, like that similar mass in the case first described, once filled up the central hollow of a crater, and that its flanks, or sloping walls, have since been worn quite away by the sea, in which it stands exposed.

Small basaltic craters.—A bare, undulating tract, at the eastern end of Chatham Island, is remarkable from the number, proximity, and form of the small basaltic craters with which it is studded. They consist, either of a mere conical pile, or, but less commonly, of a circle, of black and red, glossy scoriæ, partially cemented together. They vary in diameter from 30 to 150 yards, and rise from about 50 to 100 feet above the level of the surrounding plain. From one small eminence, I counted sixty of these craters, all of which were within a third of a mile from each other, and many were much closer. I measured the distance between two very small

craters, and found that it was only thirty yards from the
summit-rim of one, to the rim of the other. Small streams
of black, basaltic lava, containing olivine and much glassy
feldspar, have flowed from many, but not from all of these
craters. The surfaces of the more recent streams were exceedingly rugged, and were crossed by great fissures; the
older streams were only a little less rugged ; and they were
all blended and mingled together in complete confusion. The
different growth, however, of the trees on the streams, often
plainly marked their different ages. Had it not been for
this latter character, the streams could in few cases have
been distinguished ; and consequently, this wide undulatory
tract might have, (as probably many tracts have,) been erro-
neously considered as formed by one great deluge of lava,
instead of by a multitude of small streams, erupted from
many small orifices.

In several parts of this tract, and especially at the base of
the small craters, there are circular pits, with perpendicular
sides, from twenty to forty feet deep. At the foot of one
small crater, there were three of these pits. They have
probably been formed, by the falling in of the roofs of small
caverns.* In other parts, there are mammiform hillocks,
which resemble great bubbles of lava, with their summits
fissured by irregular cracks, which appeared, upon entering
them, to be very deep; lava has not flowed from these
hillocks. There are, also, other very regular, mammiform
hillocks, composed of stratified lava, and surmounted by
circular, steep-sided hollows, which, I suppose have been
formed by a body of gas, first, arching the strata into one of
the bubble-like hillocks, and then, blowing off its summit.
These several kinds of hillocks and pits, as well as the
numerous, small, scoriaceous craters, all show that this tract
has been penetrated, almost like a sieve, by the passage of

* M. Elie de Beaumont has described (Mem. pour servir, &c., tom.
iv. p. 113) many "petits cirques d'éboulement" on Etna, of some of
which the origin is historically known.

heated vapours. The more regular hillocks could only have been heaved up, whilst the lava was in a softened state.*

ALBEMARLE ISLAND.—This island consists of five, great, flat-topped craters, which, together with the one on the adjoining island of Narborough, singularly resemble each other, in form and height. The southern one is 4700 feet high, two others are 3720 feet, a third only 50 feet higher, and the remaining ones apparently of nearly the same height. Three of these are situated on one line, and their craters appear elongated in nearly the same direction. The northern crater, which is not the largest, was found by the triangulation to measure externally, no less than three miles and one-eighth of a mile, in diameter. Over the lips of these great, broad caldrons, and from little orifices near their summits, deluges of black lava have flowed down their naked sides.

Fluidity of different lavas.—Near Tagus or Banks' Cove, I examined one of these great streams of lava, which is remarkable from the evidence of its former high degree of fluidity, especially when its composition is considered. Near the sea-coast this stream is several miles in width. It consists of a black, compact base, easily fusible into a black bead, with angular and not very numerous air-cells, and thickly studded with large, fractured crystals of glassy albite,† varying from the tenth of an inch to half-an-inch, in

---

* Sir G. Mackenzie (Travels in Iceland, p. 389 to 392) has described a plain of lava at the foot of Hecla, everywhere heaved up into great bubbles or blisters. Sir George states that this cavernous lava composes the uppermost stratum; and the same fact is affirmed by Von Buch (Descript. des Isles Canaries, p. 159), with respect to the basaltic stream near Rialejo, in Teneriffe. It appears singular that it should be the upper streams that are chiefly cavernous, for one sees no reason why the upper and lower should not have been equally affected at different times;—have the inferior streams flowed beneath the pressure of the sea, and thus been flattened, after the passage through them, of bodies of gas?

† In the Cordillera of Chile, I have seen lava very closely resembling this variety at the Galapagos Archipelago. It contained, however, besides the albite, well-formed crystals of augite, and the base (perhaps in consequence of the aggregation of the augitic particles) was a shade

diameter. This lava, although at first sight appearing eminently porphyritic, cannot properly be considered so, for the crystals have evidently been enveloped, rounded, and penetrated by the lava, like fragments of foreign rock in a trap-dike. This was very clear in some specimens of a similar lava, from Abingdon Island, in which the only difference was, that the vesicles were spherical and more numerous. The albite in these lavas is in a similar condition with the leucite of Vesuvius, and with the olivine, described by Von Buch,* as projecting in great balls from the basalt of Lanzarote. Besides the albite, this lava contains scattered grains of a green mineral, with no distinct cleavage, and closely resembling olivine;† but as it fuses easily into a green glass, it belongs probably to the augitic family: at James Island, however, a similar lava contained true olivine. I obtained specimens from the actual surface, and from a depth of four feet, but they differed in no respect. The high degree of fluidity of this lava-stream was at once evident, from its smooth and gently sloping surface, from the manner in which the main stream was divided by small inequalities into little rills, and especially from the manner in which its edges, far below its source, and where it must have been in some degree cooled, thinned out to almost nothing; the actual margin consisting of loose fragments, few of which were larger than a man's head. The contrast between this margin, and the steep walls, above twenty feet high, bounding many of the basaltic streams at Ascension, is very remark-

lighter in colour. I may here remark, that in all these cases, I call the feldspathic crystals, *albite*, from their cleavage-planes (as measured by the reflecting goniometer) corresponding with those of that mineral. As, however, other species of this genus have lately been discovered to cleave in nearly the same planes with albite, this determination must be considered as only provisional. I examined the crystals in the lavas of many different parts of the Galapagos group, and I found that none of them, with the exception of some crystals from one part of James Island, cleaved in the direction of orthite or potash-feldspar.

   * Description des Isles Canaries, p. 295.

   † Humboldt mentions that he mistook a green augitic mineral, occurring in the volcanic rocks of the Cordillera of Quito, for olivine.

able. It has generally been supposed that lavas abounding
with large crystals, and including angular vesicles,* have
possessed little fluidity; but we see that the case has been
very different at Albemarle Island. The degree of fluidity
in different lavas, does not seem to correspond with any
apparent corresponding amount of difference in their composition : at Chatham Island, some streams, containing
much glassy albite and some olivine, are so rugged, that
they may be compared to a sea, frozen during a storm; whilst
the great stream at Albemarle Island, is almost as smooth
as a lake, when ruffled by a breeze. At James Island,
black basaltic lava, abounding with small grains of olivine,
presents an intermediate degree of roughness; its surface
being glossy, and the detached fragments resembling in a
very singular manner, folds of drapery, cables, and pieces of
the bark of trees.†

Craters of tuff.—About a mile southward of Banks' Cove,
there is a fine elliptic crater, about 500 feet in depth, and
three quarters of a mile in diameter. Its bottom is occupied
by a lake of brine, out of which some little crateriform hills

* The irregular and angular form of the vesicles, is probably caused
by the unequal yielding of a mass composed, in almost equal proportion,
of solid crystals and of a viscid base. It certainly seems a general circumstance, as might have been expected, that in lava, which has
possessed a high degree of fluidity, *as well as an even-sized grain*, the
vesicles are internally smooth and spherical.

† A specimen of basaltic lava, with a few small broken crystals of
albite, given me by one of the officers, is perhaps worthy of description.
It consists of cylindrical ramifications, some of which are only the
twentieth of an inch in diameter, and are drawn out into the sharpest
points. The mass has not been formed like a stalactite, for the points
terminate both upwards and downwards. Globules, only the fortieth of
an inch in diameter, have dropped from some of the points, and adhere
to the adjoining branches. The lava is vesicular, but the vesicles never
reach the surface of the branches, which are smooth and glossy. As it
is generally supposed that vesicles are always elongated in the direction
of the movement of the fluid mass, I may observe, that in these cylindrical branches, which vary from a quarter to only the twentieth of an
inch in diameter, every air-cell is spherical.

of tuff rise. The lower beds are formed of compact tuff, appearing like a subaqueous deposit; whilst the upper beds, round the entire circumference, consist of a harsh, friable tuff, of little specific gravity, but often containing fragments of rock in layers. This upper tuff contains numerous pisolitic balls, about the size of small bullets, which differ from the surrounding matter, only in being slightly harder and finer grained. The beds dip away very regularly on all sides, at angles varying, as I found by measurement, from 25 to 30 degrees. The external surface of the crater slopes at a nearly similar inclination; and is formed by slightly convex ribs, like those on the shell of a pecten or scallop, which become broader as they extend from the mouth of the crater to its base. These ribs are generally from eight to twenty feet in breadth, but sometimes they are as much as forty feet broad; and they resemble old, plastered, much flattened vaults, with the plaster scaling off in plates: they are separated from each other by gullies, deepened by alluvial action. At their upper and narrow ends, near the mouth of the crater, these ribs often consist of real hollow passages, like, but rather smaller than, those often formed by the cooling of the crust of a lava-stream, whilst the inner parts have flowed onward;—of which structure I saw many examples at Chatham Island. There can be no doubt, but that these hollow ribs or vaults have been formed in a similar manner, namely, by the setting or hardening of a superficial crust on streams of mud, which have flowed down from the upper part of the crater. In another part of this same crater, I saw open concave gutters, between one and two feet wide, which appeared to have been formed by the hardening of the lower surface of a mud-stream, instead of, as in the former case, of the upper surface. From these facts, I think, it is certain, that the tuff must have flowed as mud.* This mud may

* This conclusion is of some interest, because M. Dufrénoy (Mem. pour servir, tom. iv. p. 274) has argued from strata of tuff, apparently of similar composition with that here described, being inclined at angles between 18° and 20°, that Monte Nuevo and some other craters

have been formed either within the crater, or from ashes deposited on its upper parts, and afterwards washed down by torrents of rain. The former method, in most of the cases, appears the more probable one; at James Island, however, some beds of the friable kind of tuff, extend so continuously over an uneven surface, that probably they were formed by the falling of showers of ashes.

Within this same crater, strata of coarse tuff, chiefly composed of fragments of lava, abut, like a consolidated talus, against the inside walls. They rise to a height of between 100 and 150 feet, above the surface of the internal brine-lake; they dip inwards, and are inclined at an angle varying from 30 to 36 degrees. They appear to have been formed beneath water, probably at a period when the sea occupied the hollow of the crater. I was surprised to observe, that beds having this great inclination, did not, as far as they could be followed, thicken towards their lower extremities.

Banks' Cove.—This harbour occupies part of the interior

No. 13.

A sectional sketch of the headlands forming BANKS' COVE, showing the diverging crateriform strata, and the converging stratified talus. The highest point of these hills is 817 feet above the sea.

of Southern Italy, have been formed by upheaval. From the facts given above, of the vaulted character of the separate rills, and from the tuff not extending in horizontal sheets round these crateriform hills, no one will suppose that the strata have here been produced by elevation; and yet we see that their inclination is above 20°, and often as much as 30°. The consolidated strata, also, of the internal talus, as will be immediately seen, dips at an angle of above 30°.

of a shattered crater of tuff, larger than that last described.
All the tuff is compact, and includes numerous fragments of
lava; it appears like a subaqueous deposit. The most remarkable feature in this crater, is the great development of
strata, converging inwards, as in the last case, at a considerable inclination, and often deposited in irregular, curved
layers. These interior, converging beds, as well as the
proper, diverging, crateriform strata, are represented in the
foregoing rude, sectional sketch of the headlands, forming this Cove. The internal and external strata differ little
in composition, and the former have evidently resulted from
the wear and tear, and redeposition, of the matter forming
the external, crateriform strata. From the great development of these inner beds, a person walking round the rim of
this crater, might fancy himself on a circular, anti-clinal
ridge, of stratified sandstone and conglomerate. The sea is
wearing away the inner and outer strata, and especially the
latter; so that the inwardly converging strata, will perhaps
in some future age, be left standing alone,—a case which
might at first perplex a geologist.*

JAMES ISLAND.—Two craters of tuff on this island, are the
only remaining ones which require any notice. One of them
lies a mile and a-half inland from Puerto Grande: it is circular, about the third of a mile in diameter, and 400 feet in
depth. It differs from all the other tuff-craters which I
examined, in having the lower part of its cavity, to the
height of between 100 and 150 feet, formed by a precipitous
wall of basalt, giving to the crater the appearance of having
burst through a solid sheet of rock. The upper part of this
crater consists of strata of the altered tuff, with a semi-

* I believe that this case actually occurs in the Azores, where Dr.
Webster (Description, p. 185) has described a basin-formed, little island,
composed of *strata of tuff*, dipping inwards and bounded externally by
steep sea-worn cliffs. Dr. Daubeny supposes (on Volcanos, p. 266),
that this cavity must have been formed by a circular subsidence.
It appears to me far more probable, that we here have strata, which
were originally deposited within the hollow of a crater, of which the
exterior walls have since been removed by the sea.

resinous fracture.   Its bottom is occupied by a shallow lake
of brine, covering layers of salt, which rest on deep, black
mud.   The other crater lies at the distance of a few miles,
and is only remarkable from its size and perfect condition.
Its summit is 1200 feet above the level of the sea, and the
interior hollow is 600 feet deep.   Its external, sloping surface presented a curious appearance, from the smoothness of
the wide layers of tuff, which resembled a vast plastered
floor.   Brattle Island is, I believe, the largest crater in the
Archipelago, composed of tuff; its interior diameter is
nearly a nautical mile.   At present, it is in a ruined condition, consisting of little more than half a circle, open to
the south; its great size is probably due, in part, to internal
degradation, from the action of the sea.

Segment of a small basaltic crater.—One side of Fresh-water
Bay, in James Island, is bounded by a promontory, which
forms the last wreck of a great crater.   On the beach of
this promontory, a quadrant-shaped segment of a small,
subordinate point of eruption stands exposed.   It consists
of nine, separate, little streams of lava, piled upon each other;
and of an irregular pinnacle, about fifteen feet high, of reddish-brown, vesicular basalt, abounding with large crystals of
glassy albite, and with fused augite.   This pinnacle, and
some adjoining paps of rock on the beach, represent the

No. 14.

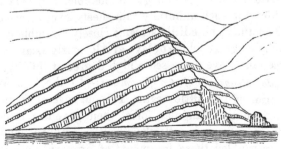

Segment of a very small orifice of eruption,
on the beach of FRESH-WATER BAY.

axis of the crater.   The streams of lava can be followed up
a little ravine, at right angles to the coast, for between ten
and fifteen yards, where they are hidden by detritus: along
the beach they are visible for nearly eighty yards, and I do
not believe that they extend much further.   The three lower
streams are united to the pinnacle; and at the point of
junction, (as is shown in the accompanying rude sketch, made
on the spot), they are slightly arched, as if in the act of
flowing over the lip of the crater.   The six upper streams,
no doubt, were originally united to this same column, before
it was worn down by the sea.   The lava of these streams is
of similar composition with that of the pinnacle, excepting
that the crystals of albite appear to be more comminuted,
and the grains of fused augite are absent.   Each stream is
separated from the one above it, by a few inches, or at most
by one or two feet in thickness, of loose, fragmentary scoriæ,
apparently derived from the abrasion of the streams, in passing over each other.   All these streams are very remarkable,
from their thinness.   I carefully measured several of them;
one was eight inches thick, but was firmly coated with three
inches above, and three inches below, of red scoriaceous rock,
(which is the case with all the streams), making altogether a
thickness of fourteen inches: this thickness was preserved
quite uniformly, along the entire length of the section.   A
second stream was only eight inches thick, including both
the upper and lower scoriaceous surfaces.   Until examining
this section, I had not thought it possible, that lava could
have flowed in such uniformly thin sheets, over a surface far
from smooth.   These little streams closely resemble in composition, that great deluge of lava at Albemarle Island, which
likewise must have possessed a high degree of fluidity.

Pseudo-extraneous, ejected fragments.—In the lava and in
the scoriæ of this little crater, I found several fragments,
which, from their angular form, their granular structure,
their freedom from air-cells, their brittle and burnt condition,
closely resembled those fragments of primary rocks, which
are occasionally ejected, as at Ascension, from volcanos.

These fragments consist of glassy albite, much mackled, and with very imperfect cleavages, mingled with semi-rounded grains, having tarnished, glossy surfaces, of a steel-blue mineral. The crystals of albite are coated by a red oxide of iron, appearing like a residual substance; and their cleavage-planes, also, are sometimes separated by excessively fine layers of this oxide, giving to the crystals the appearance of being ruled, like a glass micrometer. There was no quartz. The steel-blue mineral, which is abundant in the pinnacle, but which disappears in the streams derived from the pinnacle, has a fused appearance, and rarely presents even a trace of cleavage; I obtained, however, one measurement, which proved that it was augite; and in one other fragment, which differed from the others, in being slightly cellular and in gradually blending into the surrounding matrix, the small grains of this mineral were tolerably well crystallized. Although there is so wide a difference in appearance, between the lava of the little streams, and especially of their red scoriaceous crusts, and one of these angular, ejected fragments, which at first sight might readily be mistaken for syenite, yet I believe, that the lava has originated from the melting and movement of a mass of rock, of absolutely similar composition with the fragments. Besides the specimen above alluded to, in which we see a fragment becoming slightly cellular, and blending into the surrounding matrix, some of the grains of the steel-blue augite, also, have their surfaces becoming very finely vesicular, and passing into the nature of the surrounding paste; other grains are throughout, in an intermediate condition. The paste seems to consist of the augite more perfectly fused, or, more probably, merely disturbed in its softened state by the movement of the mass, and mingled with the oxide of iron and with finely comminuted, glassy albite. Hence probably it is, that the fused augite, which is abundant in the pinnacle, disappears in the streams. The albite is in exactly the same state, with the exception of most of the crystals being smaller, in the lava and in the embedded fragments; but in the fragments, they appear to be less abundant: this, however, would naturally happen, from the intumescence of the augitic base, and its consequent, apparent increase in bulk. It is interesting thus to trace the steps, by which a compact, granular rock becomes converted into a vesicular, pseudo-porphyritic lava, and finally into red scoriæ. The structure and composition of the embedded fragments, show that they are parts, either of a mass of primary rock, which has undergone considerable change from volcanic action, or more probably of the crust of a body of cooled and crystallized lava, which has afterwards been broken up and re-liquefied; the crust being less acted on by the renewed heat and movement.

Concluding remarks on the tuff-craters.—These craters, from the peculiarity of the resin-like substance which enters largely into their composition, from their structure, their size and number, present the most striking feature in the geology of this Archipelago. The majority of them, form either separate islets, or promontories attached to the larger islands; and those which now stand at some little distance from the coast, are worn and breached, as if by the action of the sea. From this general circumstance of their position, and from the small quantity of ejected ashes in any part of the Archipelago, I am led to conclude, that the tuff has been chiefly produced, by the grinding together of fragments of lava within active craters, communicating with the sea. In the origin and composition of the tuff, and in the frequent presence of a central lake of brine and of layers of salt, these craters resemble, though on a gigantic scale, the " salses," or hillocks of mud, which are common in some parts of Italy and in other countries.* Their closer connection, however, in this Archipelago, with ordinary volcanic action,

* D'Aubuisson's Traité de Géognosie, tom. i. p. 189. I may remark that I saw at Terceira, in the Azores, a crater of tuff or peperino, very similar to these of the Galapagos Archipelago. From the description given in Freycinet's Voyage, similar ones occur at the Sandwich Islands; and probably they are present in many other places.

is shown by the pools of solidified basalt, with which they are sometimes filled up.

It at first appears very singular, that all the craters formed of tuff have their southern sides, either quite broken down and wholly removed, or much lower than the other sides. I saw and received accounts of twenty-eight of these craters; of these, twelve form separate islets,* and now exist as mere crescents quite open to the south, with occasionally a few points of rock marking their former circumference; of the remaining sixteen, some form promontories, and others stand at a little distance inland from the shore; but all, have their southern sides either the lowest, or quite broken down. Two, however, of the sixteen, had their northern sides also low, whilst their eastern and western sides were perfect. I did not see, or hear of, a single exception to the rule, of these craters being broken down or low on the side, which faces a point of the horizon between S.E. and S.W. This rule does not apply to craters composed of lava and scoriæ. The explanation is simple: at this Archipelago, the waves from the trade-wind, and the swell propagated from the distant parts of the open ocean, coincide in direction, (which is not the case in many parts of the Pacific,) and with their united forces attack the southern sides of all the islands; and consequently the southern slope, even when entirely formed of hard basaltic rock, is invariably steeper than the northern slope. As the tuff-craters are composed of a soft material, and as probably all, or nearly all, have at some period stood immersed in the sea, we need not wonder, that they should invariably exhibit on their exposed sides, the effects of this great denuding power. Judging from the worn condition of many of these craters, it is probable, that

* These consist of the three Crossman Islets, the largest of which is 600 feet in height; Enchanted Island; Gardner Island (760 feet high); Champion Island (331 feet high); Enderby Island; Brattle Island; two islets near Indefatigable Island; and one near James Island. A second crater near James Island (with a salt-lake in its centre) has its southern side only about twenty feet high, whilst the other parts of the circumference are about 300 feet in height.

some have been entirely washed away. As there is no reason to suppose, that the craters formed of scoriæ and lava were erupted whilst standing in the sea, we can see why the rule does not apply to them. At Ascension, it was shown, that the mouths of the craters, which are there all of terrestrial origin, have been affected by the trade-wind; and this same power might here, also, aid in making the windward and exposed sides of some of the craters, originally the lowest.

Mineralogical composition of the rocks.—In the northern islands, the basaltic lavas seem generally to contain more albite than they do in the southern half of the Archipelago; but almost all the streams contain some. The albite is not unfrequently associated with olivine. I did not observe in any specimen distinguishable crystals of hornblende or augite; I except the fused grains in the ejected fragments, and in the pinnacle of the little crater, above described. I did not meet with a single specimen of true trachyte; though some of the paler lavas, when abounding with large crystals of the harsh and glassy albite, resemble in some degree this rock; but in every case the basis fuses into a black enamel. Beds of ashes and far-ejected scoriæ, as previously stated, are almost absent; nor did I see a fragment of obsidian or of pumice. Von Buch* believes that the absence of pumice on Mount Etna, is consequent on the feldspar being of the Labrador variety; if the presence of pumice depends on the constitution of the feldspar, it is remarkable, that it should be absent in this archipelago, and abundant in the Cordillera of South America, in both of which regions, the feldspar is of the albitic variety. Owing to the absence of ashes, and the general indecomposable character of the lava in this Archipelago, the islands are slowly clothed with a poor vegetation, and the scenery has a desolate and frightful aspect.

Elevation of the land.—Proofs of the rising of the land are scanty and imperfect. At Chatham Island, I noticed some great blocks of lava, cemented by calcareous matter, con-

* Description des Isles Canaries, p. 328.

taining recent shells; but they occurred at the height of only a few feet above high-water mark. One of the officers gave me some fragments of shells, which he found embedded several hundred feet above the sea, in the tuff of two craters, distant from each other. It is possible, that these fragments may have been carried up to their present height, in an eruption of mud; but as in one instance, they were associated with broken oyster-shells, almost forming a layer, it is more probable, that the tuff was uplifted with the shells, in mass. The specimens are so imperfect, that they can be recognized only as belonging to recent marine genera. On Charles Island, I observed a line of great rounded blocks, piled on the summit of a vertical cliff, at the height of fifteen feet above the line, where the sea now acts during the heaviest gales. This appeared, at first, good evidence in favour of the elevation of the land; but it was quite deceptive, for I afterwards saw on an adjoining part of this same coast, and heard from eye-witnesses, that wherever a recent stream of lava forms a smooth inclined plane, entering the sea, the waves during gales have the power of *rolling up rounded* blocks to a great height, above the line of their ordinary action. As the little cliff in the foregoing case, is formed by a stream of lava, which, before being worn back, must have entered the sea with a gently sloping surface, it is possible, or rather it is probable, that the rounded boulders, now lying on its summit, are merely the remnant of those, which had been *rolled up* during storms, to their present height.

Direction of the fissures of eruption.—The volcanic orifices in this group, cannot be considered as indiscriminately scattered. Three great craters on Albemarle Island form a well marked line, extending N.W. by N. and S.E. by S. Narborough Island, and the great crater on the rectangular projection of Albemarle Island, form a second parallel line. To the east, Hood's Island, and the islands and rocks between it and James Island, form another, nearly parallel line, which, when prolonged, includes Culpepper and Wenman Islands, lying seventy miles to the north. The other islands lying further eastward, form a less regular fourth line. Several of these islands, and the vents on Albemarle Island, are so placed, that they likewise fall on a set of rudely parallel lines, intersecting the former lines at right angles; so that the principal craters appear to lie on the points, where two sets of fissures cross each other. The islands themselves, with the exception of Albemarle Island, are not elongated in the same direction with the lines on which they stand. The direction of these islands, is nearly the same with that, which prevails in so remarkable a manner, in the numerous archipelagos of the great Pacific Ocean. Finally, I may remark, that amongst the Galapagos islands, there is no one dominant vent, much higher than all the others, as may be observed in many volcanic archipelagos : the highest, is the great mound on the south-western extremity of Albemarle Island, which exceeds by barely a thousand feet, several other neighbouring craters.

CHAPTER VI.

TRACHYTE AND BASALT.—DISTRIBUTION OF VOLCANIC ISLES.

*The sinking of crystals in fluid lava—Specific gravity of the constituent
parts of trachyte and of basalt, and their consequent separation—Obsidian—Apparent non-separation of the elements of plutonic rocks—
Origin of trap-dikes in the plutonic series—Distribution of volcanic
islands; their prevalence in the great oceans—They are generally
arranged in lines—The central volcanos of Von Buch doubtful—Volcanic islands bordering continents—Antiquity of volcanic islands, and
their elevation in mass—Eruptions on parallel lines of fissure within
the same geological period.*

*On the separation of the constituent minerals of lava, according
to their specific gravities.*—One side of Fresh-water Bay, in
James Island, is formed by the wreck of a large crater,
mentioned in the last chapter, of which the interior has been
filled up by a pool of basalt, about 200 feet in thickness.
This basalt is of a gray colour, and contains many crystals of
glassy albite, which become much more numerous in the
lower, scoriaceous part. This is contrary to what might
have been expected, for if the crystals had been originally
disseminated in equal numbers, the greater intumescence of
this lower scoriaceous part, would have made them appear
fewer in number. Von Buch* has described a stream of
obsidian on the peak of Teneriffe, in which the crystals of
feldspar become more and more numerous, as the depth or
thickness increases, so that near the lower surface of the
stream, the lava even resembles a primary rock. Von Buch
further states, that M. Drée, in his experiments in melting
lava, found that the crystals of feldspar always tended to

* Description des Isles Canaries, pp. 190 and 191.

precipitate themselves to the bottom of the crucible. In these cases, I presume there can be no doubt,* that the crystals sink from their weight. The specific gravity of feldspar varies† from 2·4 to 2·58, whilst obsidian seems commonly to be from 2·3 to 2·4; and in a fluidified state, its specific gravity would probably be less, which would facilitate the sinking of the crystals of feldspar. At James Island, the crystals of albite, although no doubt of less weight than the gray basalt, in the parts where compact, might easily be of greater specific gravity than the scoriaceous mass, formed of melted lava and bubbles of heated gas.

The sinking of crystals through a viscid substance like molten rock, as is unequivocally shown to have been the case in the experiments of M. Drée, is worthy of further consideration, as throwing light on the separation of the trachytic and basaltic series of lavas. Mr. P. Scrope has speculated on this subject; but he does not seem to have been aware of any positive facts, such as those above given; and he has overlooked one very necessary element, as it appears to me, in the phenomenon,—namely, the existence of either the lighter or heavier mineral, in globules or in crystals. In a substance of imperfect fluidity, like molten rock, it is hardly credible, that the separate, infinitely small

---

* In a mass of molten iron, it is found (Edinburgh New Philosophical Journal, vol. xxiv. p. 66) that the substances, which have a closer affinity for oxygen, than iron has, rise from the interior of the mass to the surface. But a similar cause can hardly apply to the separation of the crystals of these lava-streams. The cooling of the surface of lava seems, in some cases, to have affected its composition; for Dufrénoy (Mem. pour servir, tom. iv. p. 271) found that the interior parts of a stream near Naples, contained two-thirds of a mineral which was acted on by acids, whilst the surface consisted chiefly of a mineral unattackable by acids.

† I have taken the specific gravities of the simple minerals from Von Kobell, one of the latest and best authorities, and of the rocks from various authorities. Obsidian, according to Phillips, is 2·35; and Jameson says it never exceeds 2·4; but a specimen from Ascension, weighed by myself, was 2·42.

atoms, whether of feldspar, augite, or of any other mineral, would have power from their slightly different gravities, to overcome the friction caused by their movement; but if the atoms of any one of these minerals became, whilst the others remained fluid, united into crystals or granules, it is easy to perceive that from the lessened friction, their sinking or floating power would be greatly increased. On the other hand, if all the minerals became granulated at the same time, it is scarcely possible, from their mutual resistance, that any separation could take place. A valuable, practical discovery, illustrating the effect of the granulation of one element in a fluid mass, in aiding its separation, has lately been made; when lead containing a small proportion of silver, is constantly stirred whilst cooling, it becomes granulated, and the grains or imperfect crystals of nearly pure lead, sink to the bottom, leaving a residue of melted metal much richer in silver; whereas if the mixture be left undisturbed, although kept fluid for a length of time, the two metals show no signs of separating.* The sole use of the stirring seems to be, the formation of detached granules. The specific gravity of silver is 10·4, and of lead 11·35: the granulated lead, which sinks, is never absolutely pure, and the residual fluid metal contains, when richest, only $\frac{1}{110}$ part of silver. As the difference in specific gravity, caused by the different proportions of the two metals, is so exceedingly small, the separation is probably aided in a great degree by the difference in gravity between the lead, when granular though still hot, and when fluid.

In a body of liquefied volcanic rock, left for some time without any violent disturbance, we might expect, in accord-

* A full and interesting account of this discovery, by Mr. Pattinson, was read before the British Association in September, 1838. In some alloys, according to Turner (Chemistry, p. 210), the heaviest metal sinks, and it appears that this takes place whilst both metals are fluid. Where there is a considerable difference in gravity, as between iron and the slag formed during the fusion of the ore, we need not be surprised at the atoms separating, without either substance being granulated.

ance with the above facts, that if one of the constituent minerals became aggregated into crystals or granules, or had been enveloped in this state from some previously existing mass, such crystals or granules would rise or sink, according to their specific gravity. Now we have plain evidence of crystals being embedded in many lavas, whilst the paste or basis has continued fluid. I need only refer, as instances, to the several, great, pseudo-porphyritic streams at the Galapagos islands, and to the trachytic streams in many parts of the world, in which we find crystals of feldspar bent and broken by the movement of the surrounding, semi-fluid matter. Lavas are chiefly composed of three varieties of feldspar, varying in specific gravity from 2·4 to 2·74; of hornblende and augite, varying from 3·0 to 3·4: of olivine, varying from 3·3 to 3·4; and lastly, of oxides of iron, with specific gravities from 4·8 to 5·2. Hence crystals of feldspar, enveloped in a mass of liquefied, but not highly vesicular lava, would tend to rise to the upper parts; and crystals or granules of the other minerals, thus enveloped, would tend to sink. We ought not, however, to expect any perfect degree of separation in such viscid materials. Trachyte, which consists chiefly of feldspar, with some hornblende and oxide of iron, has a specific gravity of about 2·45;* whilst basalt composed chiefly of augite and feldspar, often with much iron and olivine, has a gravity of about 3·0. Accordingly we find, that where both trachytic and basaltic streams have proceeded from the same orifice, the trachytic streams have generally been first erupted, owing, as we must suppose, to the molten lava of this series having accumulated in the upper parts of the volcanic focus. This order of eruption has been observed by Beudant, Scrope, and by other authors; three instances, also, have been given in this volume. As the later erup-

---

* Trachyte from Java, was found by Von Buch to be 2·47; from Auvergne, by De la Beche, it was 2·42; from Ascension, by myself, it was 2·42. Jameson and other authors give to basalt a specific gravity of 3·0; but specimens from Auvergne were found, by De la Beche, to be only 2·78; and from the Giant's Causeway, to be 2·91.

tions, however, from most volcanic mountains, burst through
their basal parts, owing to the increased height and weight
of the internal column of molten rock, we see why, in most
cases, only the lower flanks of the central, trachytic masses,
are enveloped by basaltic streams. The separation of the
ingredients of a mass of lava would, perhaps, sometimes
take place within the body of a volcanic mountain, if lofty
and of great dimensions, instead of within the underground
focus; in which case, trachytic streams might be poured
forth, almost contemporaneously, or at short recurrent intervals, from its summit, and basaltic streams from its base:
this seems to have taken place at Teneriffe.* I need only
further remark, that from violent disturbances the separation of the two series, even under otherwise favourable conditions, would naturally often be prevented, and likewise
their usual order of eruption be inverted. From the high
degree of fluidity of most basaltic lavas, these perhaps,
alone, would in many cases reach the surface.

As we have seen that crystals of feldspar, in the instance
described by Von Buch, sink in obsidian, in accordance with
their known greater specific gravity, we might expect to find
in every trachytic district, where obsidian has flowed as lava,
that it had proceeded from the upper or highest orifices.
This, according to Von Buch, holds good in a remarkable
manner, both at the Lipari Islands and on the Peak of Teneriffe; at this latter place, obsidian has never flowed from
a less height than 9,200 feet. Obsidian, also, appears to
have been erupted from the loftiest peaks of the Peruvian
Cordillera. I will only further observe, that the specific
gravity of quartz varies from 2·6 to 2·8; and therefore, that
when present in a volcanic focus, it would not tend to sink
with the basaltic bases; and this, perhaps, explains the frequent presence, and the abundance of this mineral, in the
lavas of the trachytic series, as observed in previous parts of
this volume.

* Consult Von Buch's well-known and admirable *Description Physique*
of this island, which might serve as a model of descriptive geology.

An objection to the foregoing theory, will, perhaps, be drawn, from the plutonic rocks not being separated into two evidently distinct series, of different specific gravities; although, like the volcanic, they have been liquefied. In answer, it may first be remarked, that we have no evidence of the atoms of any one of the constituent minerals in the plutonic series, having been aggregated, whilst the others remained fluid, which we have endeavoured to show is an almost necessary condition of their separation; on the contrary, the crystals have generally impressed each other with their forms.*

In the second place, the perfect tranquillity, under which it is probable that the plutonic masses, buried at profound depths, have cooled, would, most likely, be highly unfavourable to the separation of their constituent minerals; for, if the attractive force, which during the progressive cooling draws together the molecules of the different minerals, has power sufficient to keep them together, the friction between such half-formed crystals or pasty globules, would effectually prevent the heavier ones from sinking, or the lighter ones from rising. On the other hand, a small amount of disturbance, which would probably occur in most volcanic foci, and which we have seen does not prevent the separation of gra-

---

* The crystalline paste of phonolite, is frequently penetrated by long needles of hornblende; from which it appears, that the hornblende, though the more fusible mineral, has crystallized before, or at the same time with, a more refractory substance. Phonolite, as far as my observations serve, in every instance appears to be an injected rock, like those of the plutonic series; hence probably, like these latter, it has generally been cooled without repeated and violent disturbances. Those geologists who have doubted whether granite could have been formed by igneous liquefaction, because minerals of different degrees of fusibility impress each other with their forms, could not have been aware of the fact of crystallized hornblende penetrating phonolite, a rock undoubtedly of igneous origin. The viscidity, which it is now known, that both feldspar and quartz retain at a temperature much below their points of fusion, easily explains their mutual impressment. Consult on this subject Mr. Horner's paper on Bonn. Geolog. Transact. vol. iv. p. 439; and L'Institut, with respect to quartz, 1839, p. 161.

nules of lead from a mixture of molten lead and silver, or crystals of feldspar from streams of lava, by breaking and dissolving the less perfectly formed globules, would permit the more perfect and therefore unbroken crystals, to sink or rise, according to their specific gravity.

Although in plutonic rocks two distinct species, corresponding to the trachytic and basaltic series, do not exist, I much suspect, that a certain amount of separation of their constituent parts has often taken place. I suspect this from having observed, how frequently dikes of greenstone and basalt intersect widely extended formations of granite and the allied metamorphic rocks. I have never examined a district in an extensive granitic region, without discovering dikes; I may instance the numerous trap-dikes, in several districts of Brazil, Chile, and Australia, and at the Cape of Good Hope: many dikes likewise occur in the great granitic tracts of India, in the north of Europe, and in other countries. Whence, then, has the greenstone and basalt, forming these dikes, come? Are we to suppose, like some of the elder geologists, that a zone of trap is uniformly spread out beneath the granitic series, which composes, as far as we know, the foundations of the earth's crust. Is it not more probable, that these dikes have been formed by fissures penetrating into partially cooled rocks of the granitic and metamorphic series, and by their more fluid parts, consisting chiefly of hornblende, oozing out, and being sucked into such fissures? At Bahia, in Brazil, in a district composed of gneiss and primitive greenstone, I saw many dikes, of a dark augitic (for one crystal certainly was of this mineral) or hornblendic rock, which, as several appearances clearly proved, either had been formed before the surrounding mass had become solid, or had together with it been afterwards thoroughly softened.* On both sides of one of

* Portions of these dikes have been broken off, and are now surrounded by the primary rocks, with their laminæ conformably winding round them. Dr. Hubbard, also, (Silliman's Journal, vol. xxxiv. p. 119), has described an interlacement of trap-veins in the granite of the

these dikes, the gneiss was penetrated to the distance of several yards, by numerous, curvilinear threads or streaks of dark matter, which resembled in form clouds of the class called cirrhi-comæ; some few of these threads could be traced to their junction with the dike. When examining them, I doubted whether such hair-like and curvilinear veins could have been injected, and I now suspect, that instead of having been injected from the dike, they were its feeders. If the foregoing view of the origin of trap-dikes in widely extended granitic regions, far from rocks of any other formation, be admitted as probable, we may further admit, in the case of a great body of plutonic rock, being impelled by repeated movements into the axis of a mountain-chain, that its more liquid constituent parts might drain into deep and unseen abysses; afterwards, perhaps, to be brought to the surface under the form, either of injected masses of greenstone and augitic porphyry,* or of basaltic eruptions. Much of the difficulty which geologists have experienced, when they have compared the composition of volcanic with plutonic formations, will, I think, be removed, if we may believe, that most plutonic masses have been, to a certain extent, drained of those comparatively weighty and easily liquefied elements, which compose the trappean and basaltic series of rocks.

On the distribution of volcanic islands.—During my investi-

White Mountains, which he thinks must have been formed when both rocks were soft.

* Mr. Phillips (Lardner's Encyclop. vol. ii. p. 115) quotes Von Buch's statement, that augitic porphyry ranges parallel to, and is found constantly at the base of, great chains of mountains. Humboldt, also, has remarked the frequent occurrence of trap-rock, in a similar position; of which fact I have observed many examples at the foot of the Chilian Cordillera. The existence of granite in the axes of great mountain chains is always probable, and I am tempted to suppose, that the laterally injected masses of augitic porphyry and of trap, bear nearly the same relation to the granitic axes, which basaltic lavas bear to the central trachytic masses, round the flanks of which they have so frequently been erupted.

gations on coral-reefs, I had occasion to consult the works of many voyagers, and I was invariably struck with the fact, that with rare exceptions, the innumerable islands scattered throughout the Pacific, Indian, and Atlantic Oceans, were composed either of volcanic, or of modern coral-rocks. It would be tedious to give a long catalogue of all the volcanic islands; but the exceptions which I have found, are easily enumerated : in the Atlantic, we have St. Paul's Rock, described in this volume, and the Falkland Islands, composed of quartz and clayslate; but these latter islands are of considerable size, and lie not very far from the South American coast :* in the Indian Ocean, the Seychelles (situated in a line prolonged from Madagascar) consist of granite and quartz : in the Pacific Ocean, New Caledonia, an island of large size, belongs (as far as is known) to the primary class; New Zealand, which contains much volcanic rock and some active volcanos, from its size cannot be classed with the small islands, which we are now considering. The presence of a small quantity of non-volcanic rock, as of clay-slate on three of the Azores,† or of tertiary limestone at Madeira, or of clay-slate at Chatham Island in the Pacific, or of lignite at Kerguelen Land, ought not to exclude such islands or archipelagos, if formed chiefly of erupted matter, from the volcanic class.

The composition of the numerous islands, scattered through the great oceans, being with such rare exceptions volcanic, is evidently an extension of that law, and the effect

* Judging from Forster's imperfect observation, perhaps Georgia is not volcanic. Dr. Allan is my informant with regard to the Seychelles. I do not know of what formation Rodriguez, in the Indian Ocean, is composed.

† This is stated on the authority of Count V. de Bedemar, with respect to Flores and Graciosa (Charlsworth Magazine of Nat. Hist. vol. i. p. 557). St. Maria has no volcanic rock, according to Captain Boyd (Von Buch's Descript.' p. 365). Chatham Island has been described by Dr. Dieffenbach, in the Geographical Journal, 1841, p. 201. As yet we have received only imperfect notices on Kerguelen Land, from the Antarctic Expedition.

of those same causes, whether chemical or mechanical, from which it results, that a vast majority of the volcanos now in action, stand either as islands in the sea, or near its shores. This fact of the ocean-islands being so generally volcanic, is, also, interesting, in relation to the nature of the mountain-chains on our continents, which are comparatively seldom volcanic ; and yet we are led to suppose, that where our continents now stand, an ocean once extended.   Do volcanic eruptions, we may ask, reach the surface more readily through fissures, formed during the first stages of the conversion of the bed of the ocean into a tract of land ?

Looking at the charts of the numerous volcanic archipelagos, we see that the islands are generally arranged, either in single, double, or treble rows, in lines which are frequently curved in a slight degree.*   Each separate island is either rounded, or more generally elongated in the same direction with the group in which it stands, but sometimes transversely to it.   Some of the groups which are not much elongated, present little symmetry in their forms; M. Virlet† states that this is the case with the Grecian Archipelago : in such groups I suspect, (for I am aware how easy it is to deceive oneself on these points), that the vents are generally arranged on one line, or on a set of short parallel lines, intersecting at nearly right angles another line, or set of lines.   The Galapagos Archipelago offers an example of this structure, for most of the islands and the chief orifices on the largest island, are so grouped as to fall on a set of lines ranging about N.W. by N., and on another set ranging about W.S.W.: in the Canary Archipelago, we have a simpler structure of the same kind : in the Cape de Verde group, which appears to be the least symmetrical of any oceanic, volcanic archipelago, a N.W. and S.E. line formed by several

* Professors William and Henry Darwin Rogers have lately insisted much, in a memoir read before the American Association, on the regularly curved lines of elevation in parts of the Appalachian range.

† Bulletin de la Soc. Géolog. tom. iii. p. 110.

islands, if prolonged, would intersect at right angles a curved line, on which the remaining islands are placed.

Von Buch* has classed all volcanos under two heads, namely, *central volcanos*, round which numerous eruptions have taken place on all sides, in a manner almost regular, and *volcanic chains*. In the examples given of the first class, as far as position is concerned, I can see no grounds for their being called 'central'; and the evidence of any difference in mineralogical nature, between *central volcanos* and *volcanic chains*, appears slight. No doubt some one island in most small volcanic archipelagos, is apt to be considerably higher than the others; in a similar manner, whatever the cause may be, that on the same island, one vent is generally higher than all the others. Von Buch does not include in his class of volcanic chains, small archipelagos, in which the islands are admitted by him, as at the Azores, to be arranged in lines; but when viewing on a map of the world, how perfect a series exists, from a few volcanic islands placed in a row, to a train of linear archipelagos following each other in a straight line, and so on to a great wall like the Cordillera of America, it is difficult to believe, that there exists any essential difference between short and long volcanic chains. Von Buch states† that his volcanic chains surmount, or are closely connected with, mountain-ranges of primary formation: but if trains of linear archipelagos are in the course of time, by the long continued action of the elevatory and volcanic forces, converted into mountain-ranges, it would naturally result, that the inferior primary rocks would often be uplifted and brought into view.

Some authors have remarked, that volcanic islands occur scattered though at very unequal distances, along the shores of the great continents, as if in some measure connected with them. In the case of Juan Fernandez, situated 330 miles from the coast of Chile, there was undoubtedly a

* Description des Isles Canaries, p. 324.
† Idem, p. 393.

connexion between the volcanic forces acting under this island, and under the continent, as was shown during the earthquake of 1835. The islands, moreover, of some of the small volcanic groups, which thus border continents, are placed in lines, related to those, along which the adjoining shores of the continents trend; I may instance the lines of intersection at the Galapagos, and at the Cape de Verde Archipelagos, and the best marked line of the Canary Islands. If these facts be not merely accidental, we see that many scattered volcanic islands and small groups are related, not only by proximity, but in the direction of the fissures of eruption to the neighbouring continents,—a relation, which Von Buch considers, characteristic of his great volcanic chains.

In volcanic archipelagos, the orifices are seldom in activity on more than one island at a time; and the greater eruptions usually recur only after long intervals. Observing the number of craters, that are usually found on each island of a group, and the vast amount of matter which has been erupted from them, one is led to attribute a high antiquity even to those groups, which appear, like the Galapagos, to be of comparatively recent origin. This conclusion accords with the prodigious amount of degradation, by the slow action of the sea, which their originally sloping coasts must have suffered, when they are worn back, as is so often the case, into grand precipices. We ought not, however, to suppose, in hardly any instance, that the whole body of matter, forming a volcanic island, has been erupted at the level, on which it now stands: the number of dikes, which seem invariably to intersect the interior parts of every volcano, show, on the principles explained by M. Elie de Beaumont, that the whole mass has been uplifted and fissured. A connexion, moreover, between volcanic eruptions and contemporaneous elevations in mass* has, I think, been shown to exist, in my work on Coral Reefs, both from the frequent presence of

---

* A similar conclusion is forced on us, by the phenomena, which accompanied the earthquake of 1835, at Conception, and which are detailed in my paper (vol. v. p. 601) in the Geological Transactions.

upraised organic remains, and from the structure of the accompanying coral-reefs. Finally, I may remark, that in the same Archipelago, eruptions have taken place within the historical period, on more than one of the parallel lines of fissure: thus at the Galapagos Archipelago, eruptions have taken place from a vent on Narborough Island, and from one on Albemarle Island, which vents do not fall on the same line; at the Canary Islands, eruptions have taken place in Teneriffe and Lanzarote; and at the Azores, on the three parallel lines of Pico, St. Jorge and Terceira. Believing that a mountain-axis differs essentially from a volcano, only in plutonic rocks having been injected, instead of volcanic matter having been ejected, this appears to me an interesting circumstance; for we may infer from it as probable, that in the elevation of a mountain-chain, two or more of the parallel lines forming it, may be upraised and injected within the same geological period.

CHAPTER VII.

New South Wales—Sandstone formation—Embedded pseudo-fragments of shale—Stratification—Current-cleavage—Great valleys.—Van Diemen's Land—Palæozoic formation—Newer formation with volcanic rocks—Travertin with leaves of extinct plants—Elevation of the land.—New Zealand—King George's Sound—Superficial ferruginous beds—Superficial calcareous deposits, with casts of branches—Their origin from drifted particles of shells and corals—Their extent.—Cape of Good Hope—Junction of the granite and clay-slate—Sandstone formation.

THE *Beagle*, in her homeward voyage, touched at New Zealand, Australia, Van Diemen's Land, and the Cape of Good Hope. In order to confine the third Part of these Geological Observations to South America, I will here briefly describe all that I observed at these places, worthy of the attention of geologists.

New South Wales.—My opportunities of observation consisted of a ride of ninety geographical miles to Bathurst, in a W.N.W. direction from Sydney. The first thirty miles from the coast passes over a sandstone country, broken up in many places by trap-rocks, and separated by a bold escarpement overhanging the river Nepean, from the great sandstone platform of the Blue Mountains. This upper platform is 1000 feet high at the edge of the escarpement, and rises in a distance of 25 miles to between 3000 and 4000 feet above the level of the sea. At this distance, the road descends to a country rather less elevated, and composed in chief part of primary rocks. There is much granite, in one part passing into a red porphyry with octagonal crystals of quartz, and intersected in some places by trap-dikes. Near the Downs of Bathurst, I passed over much pale-brown, glossy clay-slate, with the shattered laminæ running north and south: I mention this fact, because Captain King informs me, that in the country a hundred miles southward, near Lake George, the mica-slate ranges so invariably north and south, that the inhabitants take advantage of it in finding their way through the forests.

The sandstone of the Blue Mountains is at least 1,200 feet thick, and in some parts is apparently of greater thickness; it consists of small grains of quartz, cemented by white earthy matter, and it abounds with ferruginous veins. The lower beds sometimes alternate with shales and coal: at Wolgan I found in carbonaceous shale, leaves of the *Glossopteris Brownii*, a fern which so frequently accompanies the coal of Australia. The sandstone contains pebbles of quartz; and these generally increase in number and size (seldom, however, exceeding an inch or two in diameter) in the upper beds: I observed a similar circumstance in the grand sandstone formation at the Cape of Good Hope. On the South American coast, where tertiary and supra-tertiary beds have been extensively elevated, I repeatedly noticed that the uppermost beds were formed of coarser materials than the lower: this appears to indicate that, as the sea became shallower, the force of the waves or currents increased. On the lower platform, however, between the Blue Mountains and the coast, I observed that the upper beds of the sandstone frequently passed into argillaceous shale,—the effect probably, of this lower space having been protected from strong currents during its elevation. The sandstone of the Blue Mountains evidently having been of mechanical origin, and not having suffered any metamorphic action, I was surprised at observing, that in some specimens nearly all the grains of quartz were so perfectly crystallized with brilliant facets, that they evidently had not in their *present* form been aggregated in any previously existing

rock.* It is difficult to imagine how these crystals could
have been formed; one can hardly believe that they were
separately precipitated in their present crystallized state.
Is it possible, that rounded grains of quartz may have been
acted on by a fluid corroding their surfaces, and depositing
on them fresh silica? I may remark, that in the sandstone
formation of the Cape of Good Hope, it is evident, that
silica has been profusely deposited from aqueous solution.

In several parts of the sandstone, I noticed patches of
shale, which might at the first glance have been mistaken
for extraneous fragments; their horizontal laminæ, however, being parallel with those of the sandstone, showed that
they were the remnants of thin, continuous beds. One such
fragment (probably the section of a long narrow strip) seen
in the face of a cliff, was of greater vertical thickness than
breadth, which proves that this bed of shale must have
been in some slight degree consolidated, after having been
deposited, and before being worn away by the currents.
Each patch of the shale shows, also, how slowly many of
the successive layers of sandstone were deposited. These
pseudo-fragments of shale will perhaps explain in some
cases, the origin of apparently extraneous fragments in
crystalline metamorphic rocks. I mention this, because
I found near Rio de Janeiro a well-defined angular fragment,
seven yards long by two yards in breadth, of gneiss containing garnets and mica in layers, enclosed in the ordinary,
stratified, porphyritic gneiss of the country. The laminæ of
the fragment and of the surrounding matrix, ran in exactly
the same direction, but they dipped at different angles. I
do not wish to affirm that this singular fragment (a solitary

* I have lately seen, in a paper by Smith (the father of English
geologists), in the Magazine of Natural History, that the grains of
quartz in the mill-stone grit of England are often crystallized. Sir
David Brewster, in a paper read before the British Association, 1840,
states, that in old decomposed glass, the silex and metals separate into
concentric rings, and that the silex regains its crystalline structure, as
is shown by its action on light.

case, as far as I know) was originally deposited in a layer, like the shale in the Blue Mountains, between the strata of the porphyritic gneiss, before they were metamorphosed; but there is sufficient analogy between the two cases to render such an explanation possible.

Stratification of the escarpement.—The strata of the Blue Mountains appear to the eye horizontal; but they probably have a similar inclination with the surface of the platform, which slopes from the west towards the escarpement over the Nepean, at an angle of one degree, or of one hundred feet in a mile.* The strata of the escarpement dip almost conformably with its steeply inclined face, and with so much regularity, that they appear as if thrown into their present position; but on a more careful examination, they are seen to thicken and to thin out, and in the upper part to be succeeded and almost capped by horizontal beds. These appearances render it probable, that we here see an original escarpement, not formed by the sea having eaten back into the strata, but by the strata having originally extended only thus far. Those who have been in the habit of examining accurate charts of sea-coasts, where sediment is accumulating, will be aware, that the surfaces of the banks thus formed, generally slope from the coast very gently towards a certain line in the offing, beyond which the depth in most cases suddenly becomes great. I may instance the great banks of sediment within the West Indian Archipelago,† which terminate in submarine slopes, inclined at angles of between 30 and 40 degrees, and sometimes even at more than

---

* This is stated on the authority of Sir T. Mitchell, in his Travels, vol. ii. p. 357.

† I have described these very curious banks in the Appendix (p. 196) to my volume on the structure of Coral Reefs. I have ascertained the inclination of the edges of the banks, from information given me by Captain B. Allen, one of the surveyors, and by carefully measuring the horizontal distances between the last sounding on the bank and the first in the deep water. Widely extended banks in all parts of the West Indies, have the same general form of surface.

40 degrees: every one knows how steep such a slope would appear on the land. Banks of this nature, if uplifted, would probably have nearly the same external form as the platform of the Blue Mountains, where it abruptly terminates over the Nepean.

Current cleavage.—The strata of sandstone in the low coast country, and likewise on the Blue Mountains, are often divided by cross or current laminæ, which dip in different directions, and frequently at an angle of forty-five degrees. Most authors have attributed these cross layers to successive small accumulations on an inclined surface; but from a careful examination in some parts of the New Red sandstone of England, I believe that such layers generally form parts of a series of curves, like gigantic tidal-ripples, the tops of which have since been cut off, either by nearly horizontal layers, or by another set of great ripples, the folds of which do not exactly coincide with those below them. It is well known to surveyors that mud and sand are disturbed during storms at considerable depths, at least from 300 to 450 feet,* so that the nature of the bottom even becomes temporarily changed; the bottom, also, at a depth between 60 and 70 feet, has been observed† to be broadly rippled. One may, therefore, be allowed to suspect, from the appearances just mentioned in the New Red sandstone, that at greater depths, the bed of the ocean is heaped up during gales into great ripple-like furrows and depressions, which are afterwards cut off by the currents during more tranquil weather, and again furrowed during gales.

Valleys in the sandstone platforms.—The grand valleys, by which the Blue Mountains and the other sandstone platforms of this part of Australia are penetrated, and which long offered an insuperable obstacle to the attempts of the most enterprising colonist to reach the interior country, form the most striking feature in the geology of New South

* See Martin White, on Soundings in the British Channel, pp. 4 and 166.

† M. Siau on the Action of Waves. Edin. New Phil. Journ. vol. xxxi. p. 245.

Wales. They are of grand dimensions, and are bordered by continuous lines of lofty cliffs. It is not easy to conceive a more magnificent spectacle, than is presented to a person walking on the summit-plains, when without any notice he arrives at the brink of one of these cliffs, which are so perpendicular, that he can strike with a stone (as I have tried) the trees growing, at the depth of between 1000 and 1500 feet below him; on both hands he sees headland beyond headland of the receding line of cliff, and on the opposite side of the valley, often at the distance of several miles, he beholds another line rising up to the same height with that on which he stands, and formed of the same horizontal strata of pale sandstone. The bottom of these valleys are moderately level, and the fall of the rivers flowing in them, according to Sir T. Mitchell, is gentle. The main valleys often send into the platform great bay-like arms, which expand at their upper ends; and on the other hand, the platform often sends promontories into the valley, and even leaves in them great, almost insulated, masses. So continuous are the bounding lines of cliff, that to descend into some of these valleys, it is necessary to go round twenty miles; and into others, the surveyors have only lately penetrated, and the colonists have not yet been able to drive in their cattle. But the most remarkable point of structure in these valleys, is, that although several miles wide in their upper parts, they generally contract towards their mouths to such a degree, as to become impassable. The Surveyor-General, Sir T. Mitchell,* in vain endeavoured, first on foot and then by crawling between the great fallen fragments of sandstone, to ascend through the gorge by which the river Grose joins the Nepean; yet the valley of the Grose in its upper part, as I saw, forms a magnificent basin some miles in width, and is on all sides surrounded by cliffs, the summits of which are believed to be nowhere less than 3000 feet above the level of the sea. When cattle are driven into the valley

* Travels in Australia, vol. i. p. 154.—I must express my obligation to Sir T. Mitchell, for several interesting personal communications on the subject of these great valleys of New South Wales.

of the Wolgan by a path (which I descended) partly cut by
the colonists, they cannot escape; for this valley is in every
other part surrounded by perpendicular cliffs, and eight
miles lower down, it contracts, from an average width of half
a mile, to a mere chasm impassable to man or beast. Sir
T. Mitchell* states, that the great valley of the Cox river
with all its branches, contracts, where it unites with the
Nepean, into a gorge 2200 yards wide, and about 1000 feet
in depth. Other similar cases might have been added.

The first impression, from seeing the correspondence of
the horizontal strata, on each side of these valleys and great
amphitheatrical depressions, is that they have been in chief
part hollowed out, like other valleys, by aqueous erosion;
but when one reflects on the enormous amount of stone,
which on this view must have been removed, in most of the
above cases through mere gorges or chasms, one is led to
ask whether these spaces may not have subsided. But considering the form of the irregularly branching valleys, and
of the narrow promontories, projecting into them from the
platforms, we are compelled to abandon this notion. To
attribute these hollows to alluvial action, would be pre-
posterous; nor does the drainage from the summit-level
always fall, as I remarked near the Weatherboard, into the
head of these valleys, but into one side of their bay-like
recesses. Some of the inhabitants remarked to me, that
they never viewed one of these bay-like recesses, with the
headlands receding on both hands, without being struck
with their resemblance to a bold sea-coast. This is certainly
the case; moreover, the numerous fine harbours, with their
widely branching arms, on the present coast of New South
Wales, which are generally connected with the sea by a
narrow mouth, from one mile to a quarter of a mile in
width, passing through the sandstone coast-cliffs, present a
likeness, though on a miniature scale, to the great valleys of
the interior. But then immediately occurs the startling
difficulty, why has the sea worn out these great, though

* Idem, vol. ii. p. 358.

circumscribed, depressions on a wide platform, and left mere gorges, through which the whole vast amount of triturated matter must have been carried away? The only light I can throw on this enigma, is by showing that banks appear to be forming in some seas of the most irregular forms, and that the sides of such banks are so steep (as before stated) that a comparatively small amount of subsequent erosion would form them into cliffs: that the waves have power to form high and precipitous cliffs, even in landlocked harbours, I have observed in many parts of South America. In the Red Sea, banks with an extremely irregular outline and composed of sediment, are penetrated by the most singularly shaped creeks with narrow mouths: this is likewise the case, though on a larger scale, with the Bahama Banks. Such banks, I have been led to suppose,* have been formed by currents heaping sediment on an irregular bottom. That in some cases, the sea, instead of spreading out sediment in a uniform sheet, heaps it round submarine rocks and islands, it is hardly possible to doubt, after having examined the charts of the West Indies. To apply these ideas to the sandstone platforms of New South Wales, I imagine that the strata might have been heaped on an irregular bottom by the action of strong currents, and of the undulations of an open sea; and that the valley-like spaces thus left unfilled might, during a slow elevation of the land, have had their steeply sloping flanks worn into cliffs; the worn-down sandstone being removed, either at the time when the narrow gorges were cut by the retreating sea, or subsequently by alluvial action.

* See the Appendix (pp. 192 and 196) to the Part on Coral Reefs. The fact of the sea heaping up mud round a submarine nucleus, is worthy of the notice of geologists: for outlyers of the same composition with the coast-banks, are thus formed; and these, if upheaved and worn into cliffs, would naturally be thought to have been once connected together.

Van Diemen's Land.

The southern part of this island is mainly formed of mountains of greenstone, which often assumes a syenitic character, and contains much hypersthene. These mountains, in their lower half, are generally encased by strata containing numerous small corals and some shells. These shells have been examined by Mr. G. B. Sowerby, and are described in the Appendix; they consist of two species of Producta, and of six of Spirifera; two of these, namely, *P. rugata* and *S. rotundata*, resemble, as far as their imperfect condition allows of comparison, British mountain-limestone shells. Mr. Lonsdale has had the kindness to examine the corals; they consist of six undescribed species, belonging to three genera. Species of these genera occur in the Silurian, Devonian, and Carboniferous strata of Europe. Mr. Lonsdale remarks, that all these fossils have undoubtedly a Palæozoic character, and that probably they correspond in age to a division of the system, above the Silurian formations.

The strata containing these remains are singular from the extreme variability of their mineralogical composition. Every intermediate form is present, between flinty-slate, clay-slate passing into gray-wacke, pure limestone, sandstone, and porcellanic rock; and some of the beds can only be described, as composed of a siliceo-calcareo-clayslate. The formation, as far as I could judge, is at least a thousand feet in thickness: the upper few hundred feet usually consist of a siliceous sandstone, containing pebbles and no organic remains; the inferior strata, of which a pale flinty slate is perhaps the most abundant, are the most variable; and these chiefly abound with the remains. Between two beds of hard crystalline limestone, near Newtown, a layer of white soft calcareous matter is quarried, and is used for whitewashing houses. From information given to me by Mr. Frankland, the surveyor-general, it appears that this Palæozoic formation is found in different parts of the whole island;

from the same authority, I may add that on the north-eastern coast, and in Bass' Straits primary rocks extensively occur.

The shores of Storm Bay are skirted, to the height of a few hundred feet, by strata of sandstone, containing pebbles of the formation just described, with its characteristic fossils, and therefore belonging to a subsequent age. These strata of sandstone often pass into shale, and alternate with layers of impure coal; they have in many places been violently disturbed. Near Hobart Town, I observed one dike, nearly a hundred yards in width, on one side of which the strata were tilted at an angle of 60°, and on the other they were in some parts vertical, and had been altered by the effects of the heat. On the west side of Storm Bay, I found these strata capped by streams of basaltic lava with olivine ; and close by there was a mass of brecciated scoriæ, containing pebbles of lava, which probably marks the place of an ancient submarine crater. Two of these streams of basalt were separated from each other by a layer of argillaceous wacke, which could be traced passing into partially altered scoriæ. The wacke contained numerous rounded grains of a soft, grass-green mineral, with a waxy lustre, and translucent on its edges : under the blowpipe it instantly blackened, and the points fused into a strongly magnetic, black enamel. In these characters, it resembles those masses of decomposed olivine, described at St. Jago in the Cape de Verde group ; and I should have thought that it had thus originated, had I not found a similar substance, in cylindrical threads, within the cells of the vesicular basalt,—a state under which olivine never appears ; this substance,* I believe, would be classed as bole by mineralogists.

* Chlorophæite, described by Dr. MacCulloch (Western Islands, vol. i. p. 504) as occurring in a basaltic amygdaloid, differs from this substance, in remaining unchanged before the blow-pipe, and in blackening from exposure to the air. May we suppose that olivine, in undergoing the remarkable change described at St. Jago, passes through several states ?

Travertin with extinct plants.—Behind Hobart Town there is a small quarry of a hard travertin, the lower strata of which abound with distinct impressions of leaves. Mr. Robert Brown has had the kindness to look at my specimens; he informs me that there are four or five kinds, none of which he recognizes as belonging to existing species. The most remarkable leaf is palmate, like that of a fan-palm, and no plant having leaves of this structure has hitherto been discovered in Van Diemen's Land. The other leaves do not resemble the most usual form of the Eucalyptus, (of which tribe the existing forests are chiefly composed,) nor do they resemble that class of exceptions to the common form of the leaves of the Eucalyptus, which occur in this island. The travertin containing this remnant of a lost vegetation, is of pale yellow colour, hard, and in parts even crystalline; but not compact, and is everywhere penetrated by minute, tortuous, cylindrical pores. It contains a very few pebbles of quartz, and occasional layers of chalcedonic nodules, like those of chert in our Greensand. From the pureness of this calcareous rock, it has been searched for in other places, but has never been found. From this circumstance, and from the character of the deposit, it was probably formed by a calcareous spring entering a small pool or narrow creek. The strata have subsequently been tilted and fissured; and the surface has been covered by a singular mass, with which, also, a large fissure has been filled up, formed of balls of trap embedded in a mixture of wacke and a white, earthy, alumino-calcareous substance. Hence it would appear, as if a volcanic eruption had taken place on the borders of the pool, in which the calcareous matter was depositing, and had broken it up and drained it.

Elevation of the land.—Both the eastern and western shores of the Bay, in the neighbourhood of Hobart Town, are in most parts covered, to the height of thirty feet above the level of high-water mark, with broken shells, mingled with pebbles. The colonists attribute these shells to the aborigines having carried them up for food: undoubtedly, there are many large mounds, as was pointed out to me by Mr. Frankland, which have been thus formed; but I think from the numbers of the shells, from their frequent small size, from the manner in which they are thinly scattered, and from some appearances in the form of the land, that we must attribute the presence of the greater number to a small elevation of the land. On the shore of Ralph Bay (opening into Storm Bay) I observed a continuous beach about fifteen feet above high-water mark, clothed with vegetation, and by digging into it, pebbles encrusted with serpulæ were found: along the banks, also, of the river Derwent, I found a bed of broken sea-shells above the surface of the river, and at a point where the water is now much too fresh for sea-shells to live; but in both these cases, it is just possible, that before certain spits of sand and banks of mud in Storm Bay were accumulated, the tides might have risen to the height where we now find the shells.*

Evidence more or less distinct of a change of level be-tween the land and water, has been detected on almost all the land on this side of the globe. Capt. Grey, and other travellers, have found in southern Australia upraised shells, belonging either to the recent, or to a late tertiary period. The French naturalists in Baudin's expedition, found shells similarly circumstanced on the S.W. coast of Australia. The Rev. W. B. Clarke† finds proofs of the elevation of the land, to the amount of 400 feet, at the Cape of Good Hope. In the neighbourhood of the Bay of Islands in New Zea-

---

* It would appear that some changes are now in progress in Ralph Bay, for I was assured by an intelligent farmer, that oysters were formerly abundant in it, but that about the year 1834 they had, without any apparent cause, disappeared. In the Transactions of the Maryland Academy (vol. i. part i, p. 28) there is an account by Mr. Ducatel, of vast beds of oysters and clams having been destroyed by the gradual filling up of the shallow lagoons and channels, on the shores of the southern United States. At Chiloe, in South America, I heard of a similar loss, sustained by the inhabitants, in the disappearance from one part of the coast of an edible species of Ascidia.

† Proceedings of the Geological Society, vol. iii. p. 420.

land,* I observed that the shores were scattered to some
height, as at Van Diemen's Land, with sea-shells, which the
colonists attribute to the natives.   Whatever may have been
the origin of these shells, I cannot doubt, after having seen a
section of the valley of the Thames River (37° S.), drawn by
the Rev. W. Williams, that the land has been there elevated:
on the opposite sides of this great valley, three step-like terraces, composed of an enormous accumulation of rounded
pebbles, exactly correspond with each other: the escarpement of each terrace is about fifty feet in height.   No one
after having examined the terraces in the valleys on the
western shores of South America, which are strewed with
sea-shells, and have been formed during intervals of rest in
the slow elevation of the land, could doubt that the New
Zealand terraces have been similarly formed.   I may add,
that Dr. Dieffenbach, in his description of the Chatham
Islands,† (S.W. of New Zealand) states that it is manifest
" that the sea has left many places bare, which were once
covered by its waters."

King George's Sound.

This settlement is situated at the south-western angle of the
Australian continent: the whole country is granitic, with the

* I will here give a catalogue of the rocks which I met with near the
Bay of Islands, in New Zealand :—1st, Much basaltic lava, and scoriform rocks, forming distinct craters;—2nd, A castellated hill of horizontal strata of flesh-coloured limestone, showing when fractured distinct crystalline facets : the rain has acted on this rock in a remarkable
manner, corroding its surface into a miniature model of an Alpine
country : I observed here layers of chert and clay iron-stone; and in
the bed of a stream, pebbles of clay-slate;—3rd, The shores of the Bay
of Islands are formed of a feldspathic rock, of a blueish-gray colour, often
much decomposed, with an angular fracture, and crossed by numerous
ferruginous seams, but without any distinct stratification or cleavage.
Some varieties are highly crystalline, and would at once be pronounced
to be trap; others strikingly resembled clay-slate, slightly altered by
heat: I was unable to form any decided opinion on this formation.
   † Geographical Journal, vol. xi. pp. 202, 205.

constituent minerals sometimes obscurely arranged in straight or curved laminæ. In these cases, the rock would be called by Humboldt, gneiss-granite, and it is remarkable that the form of the bare conical hills, appearing to be composed of great folding layers, strikingly resembles, on a small scale, those composed of gneiss-granite at Rio de Janeiro, and those described by Humboldt at Venezuela. These plutonic rocks are, in many places, intersected by trappean-dikes : in one place, I found ten parallel dikes ranging in an E. and W. line ; and not far off another set of eight dikes, composed of a different variety of trap, ranging at right angles to the former ones. I have observed in several primary districts, the occurrence of systems of dikes parallel and close to each other.

Superficial ferruginous beds. — The lower parts of the country are everywhere covered by a bed, following the inequalities of the surface, of a honeycombed sandstone, abounding with oxides of iron. Beds of nearly similar composition are common, I believe, along the whole western coast of Australia, and on many of the East Indian islands. At the Cape of Good Hope, at the base of the mountains formed of granite and capped with sandstone, the ground is everywhere coated either by a fine-grained, rubbly, ochraceous mass, like that at King George's Sound, or by a coarser sandstone with fragments of quartz, and rendered hard and heavy by an abundance of the hydrate of iron, which presents, when freshly broken, a metallic lustre. Both these varieties have a very irregular texture, including spaces either rounded or angular, full of loose sand; from this cause the surface is always honey-combed. The oxide of iron is most abundant on the edges of the cavities, where alone it affords a metallic fracture. In these formations, as well as in many true sedimentary deposits, it is evident that iron tends to become aggregated, either in the form of a shell, or of a network. The origin of these superficial beds, though sufficiently obscure, seems to be due to alluvial action on detritus abounding with iron.

Superficial calcareous deposit.—A calcareous deposit on the summit of Bald Head, containing branched bodies, supposed by some authors to have been corals, has been celebrated by the descriptions of many distinguished voyagers.* It folds round and conceals irregular hummocks of granite, at the height of 600 feet above the level of the sea. It varies much in thickness; where stratified, the beds are often inclined at high angles, even as much as at 30 degrees, and they dip in all directions. These beds are sometimes crossed by oblique and even-sided laminæ. The deposit consists either of a fine, white, calcareous powder, in which not a trace of structure can be discovered, or of exceedingly minute, rounded grains, of brown, yellowish, and purplish colours; both varieties being generally, but not always, mixed with small particles of quartz, and being cemented into a more or less perfect stone. The rounded calcareous grains, when heated in a slight degree, instantly lose their colours; in this and in every other respect, closely resembling those minute, equal-sized particles of shells and corals, which at St. Helena have been drifted up the sides of the mountains, and have thus been winnowed of all coarser fragments. I cannot doubt, that the coloured calcareous particles here have had a similar origin. The impalpable powder has probably been derived from the decay of the rounded particles; this certainly is possible, for on the coast of Peru, I have traced *large unbroken* shells gradually falling into a substance, as fine as powdered chalk. Both of the above-mentioned varieties of calcareous sandstone frequently alternate with, and blend into, thin layers of a hard sub-stalagmitic† rock, which, even when the stone on each side

* I visited this hill, in company with Captain FitzRoy, and we came to a similar conclusion regarding these branching bodies.

† I adopt this term from Lieut. Nelson's excellent paper on the Bermuda Islands (Geolog. Trans. vol. v. p. 106), for the hard, compact, cream or brown-coloured stone, without any crystalline structure, which so often accompanies superficial calcareous accumulations. I have observed such superficial beds, coated with sub-stalagmitic rock, at the Cape of Good Hope, in several parts of Chile, and over wide

contains particles of quartz, is entirely free from them : hence we must suppose that these layers, as well as certain vein-like masses, have been formed by rain dissolving the calcareous matter and re-precipitating it, as has happened at St. Helena. Each layer probably marks a fresh surface, when the, now firmly cemented, particles existed as loose sand. These layers are sometimes brecciated and re-cemented, as if they had been broken by the slipping of the sand when soft. I did not find a single fragment of a sea-shell; but bleached shells of the *Helix melo*, an existing land species, abound in all the strata ; and I likewise found another Helix, and the case of an Oniscus.

The branches are absolutely undistinguishable in shape, from the broken and upright stumps of a thicket ; their roots are often uncovered, and are seen to diverge on all sides ; here and there a branch lies prostrate. The branches generally consist of the sandstone, rather firmer than the surrounding matter, with the central parts filled, either with friable calcareous matter, or with a sub-stalagmitic variety ; this central part is also frequently penetrated by linear crevices, sometimes, though rarely, containing a trace of

spaces in La Plata and Patagonia. Some of these beds have been formed from decayed shells, but the origin of the greater number is sufficiently obscure. The causes which determine water to dissolve lime, and then soon to redeposit it, are not, I think, known. The surface of the sub-stalagmitic layers appears always to be corroded by the rain-water. As all the above-mentioned countries have a long dry season, compared with the rainy one, I should have thought that the presence of the sub-stalagmite was connected with the climate, had not Lieut. Nelson found this substance forming under sea-water. Disintegrated shell seems to be extremely soluble; of which I found good evidence, in a curious rock at Coquimbo in Chile, which consisted of small, pellucid, empty husks, cemented together. A series of specimens clearly showed, that these husks had originally contained small rounded particles of shells, which had been enveloped and cemented together by calcareous matter, (as often happens on sea-beaches), and which subsequently had decayed, and been dissolved by water, that must have penetrated through the calcareous husks, without corroding them,—of which processes, every stage could be seen.

woody matter.   These calcareous, branching bodies, appear
to have been formed, by fine calcareous matter being washed
into the casts or cavities, left by the decay of branches and
roots of thickets, buried under drifted sand.   The whole surface of the hill is now undergoing disintegration, and hence
the casts, which are compact and hard, are left projecting.
In calcareous sand at the Cape of Good Hope, I found the
casts, described by Abel, quite similar to these at Bald
Head; but their centres are often filled with black carbonaceous matter, not yet removed.   It is not surprising,
that the woody matter should have been almost entirely
removed from the casts on Bald Head; for it is certain, that
many centuries must have elapsed since the thickets were
buried; at present, owing to the form and height of the
narrow promontory, no sand is drifted up, and the whole
surface, as I have remarked, is wearing away.   We must,
therefore, look back to a period when the land stood lower,
of which the French naturalists* found evidence in upraised
shells of recent species, for the drifting on Bald Head of the
calcareous and quartzose sand, and the consequent embed-
ment of the vegetable remains.   There was only one appearance which at first made me doubt concerning the
origin of the cast,—namely, that the finer roots from different stems sometimes became united together into upright
plates or veins; but when the manner is borne in mind, in
which fine roots often fill up cracks in hard earth, and that
these roots would decay and leave hollows, as well as the
stems, there is no real difficulty in this case.   Besides the
calcareous branches from the Cape of Good Hope, I have
seen casts, of exactly the same forms, from Madeira† and

* See M. Péron's Voyage, tom. i. p. 204.

† Dr. J. Macaulay has fully described (Eding. New Phil. Journ.
vol. xxix. p. 350) the casts from Madeira. He considers (differently
from Mr. Smith of Jordan Hill) these bodies to be corals, and the calcareous deposit to be of subaqueous origin. His arguments chiefly rest
(for his remarks on their structure are vague) on the great quantity of
the calcareous matter, and on the casts containing animal matter, as

from Bermuda; at this latter place, the surrounding calcareous rocks, judging from the specimens collected by
Lieut. Nelson, are likewise similar, as is their subaërial
formation. Reflecting on the stratification of the deposit on
Bald Head,—on the irregularly alternating layers of substalagmitic rock,— on the uniformly sized, and rounded
particles, apparently of sea-shells and corals, — on the abundance of land-shells throughout the mass, — and finally, on
the absolute resemblance of the calcareous casts, to the
stumps, roots, and branches of that kind of vegetation, which
would grow on sand-hillocks, I think there can be no reasonable
doubt, notwithstanding the different opinion of some authors,
that a true view of their origin has been here given.

Calcareous deposits, like these of King George's Sound,
are of vast extent on the Australian shores. Dr. Fitton
remarks, that " recent calcareous breccia (by which term all
these deposits are included) was found during Baudin's
voyage, over a space of no less than 25 degrees of latitude
and an equal extent of longitude, on the southern, western,
and north-western coasts."* It appears also from M.

shown by their evolving ammonia. Had Dr. Macaulay seen the enormous masses of rolled particles of shells and corals on the beach of
Ascension, and especially on coral-reefs; and had he reflected on the
effects of long-continued, gentle winds, in drifting up the finer particles,
he would hardly have advanced the argument of quantity, which is
seldom trustworthy in geology. If the calcareous matter has originated
from disintegrated shells and corals, the presence of animal matter is
what might have been expected. Mr. Anderson analyzed for Dr.
Macaulay part of a cast, and he found it composed of—

Carbonate of lime	73·15
Silica	11·90
Phosphate of lime	8·81
Animal matter	4·25
Sulphate of lime	a trace
	98·11

* For ample details on this formation, consult Dr. Fitton's Appendix
to Capt. King's Voyage. Dr. Fitton is inclined to attribute a concretionary origin to the branching bodies: I may remark, that I have

Péron, with whose observations and opinions on the origin of the calcareous matter and branching casts, mine entirely accord, that the deposit is generally much more continuous, than near King George's Sound. At Swan River, Archdeacon Scott* states that in one part it extends ten miles inland. Captain Wickham, moreover, informs me that during his late survey of the western coast, the bottom of the sea, wherever the vessel anchored, was ascertained by crow-bars being let down, to consist of white calcareous matter. Hence it seems that along this coast, as at Bermuda and at Keeling Atoll, submarine and subaërial deposits are contemporaneously in process of formation, from the disintegration of marine organic bodies. The extent of these deposits, considering their origin, is very striking ; and they can be compared in this respect, only with the great coral-reefs of the Indian and Pacific Oceans. In other parts of the world, for instance in South America, there are *superficial* calcareous deposits of great extent, in which not a trace of organic structure is discoverable ; these observations would lead to the enquiry, whether such deposits may not, also, have been formed from disintegrated shells and corals.

Cape of Good Hope.

After the accounts given by Barrow, Carmichael, Basil Hall, and W. B. Clarke of the geology of this district, I shall confine myself to a few observations on the junction of the three principal formations. The fundamental rock is granite,† overlaid by clay-slate : the latter is generally hard,

seen in beds of sand in La Plata, cylindrical stems, which no doubt thus originated ; but they differed much in appearance from these at Bald Head, and the other places above specified.

* Proceedings of the Geolog. Soc. vol. i. p. 320.

† In several places I observed in the granite, small dark-coloured balls, composed of minute scales of black mica in a tough basis. In another place, I found crystals of black schorl radiating from a common centre. Dr. Andrew Smith found in the interior parts of the country, some beautiful specimens of granite, with silvery mica radiating or rather branching, like moss, from central points. At the Geological

and glossy from containing minute scales of mica; it alternates with, and passes into, beds of slightly crystalline, feldspathic, slaty rock. This clay-slate is remarkable from being in some places (as on the Lion's Rump) decomposed, even to the depth of twenty feet, into a pale-coloured, sandstone-like rock, which has been mistaken, I believe, by some observers, for a separate formation. I was guided by Dr. Andrew Smith to a fine junction at Green Point between the granite and clay-slate: the latter at the distance of a quarter of a mile from the spot, where the granite appears on the beach, (though, probably, the granite is much nearer under-ground), becomes slightly more compact and crystalline. At a less distance, some of the beds of clay-slate are of a homogeneous texture, and obscurely striped with different zones of colour, whilst others are obscurely spotted. Within a hundred yards of the first vein of granite, the clay-slate consists of several varieties; some compact with a tinge of purple, others glistening with numerous minute scales of mica and imperfectly crystallized feldspar; some obscurely granular, others porphyritic with small, elongated spots of a soft white mineral, which being easily corroded, gives to this variety a vesicular appearance. Close to the granite, the clay-slate is changed into a dark-coloured, laminated rock, having a granular fracture, which is due to imperfect crystals of feldspar, coated by minute, brilliant, scales of mica.

The actual junction between the granitic and clay-slate districts, extends over a width of about 200 yards, and consists of irregular masses and of numerous dikes of granite, entangled and surrounded by the clay-slate: most of the dikes range in a N.W. and S.E. line, parallel to the cleavage of the slate. As we leave the junction, thin beds, and lastly, mere films of the altered clay-slate are seen, quite isolated, as if floating, in the coarsely-crystallized granite; but although completely detached, they all retain traces of the uniform

Society, there are specimens of granite with crystallized feldspar branching and radiating in like manner.

N.W. and S.E. cleavage. This fact has been observed in other similar cases, and has been advanced by some eminent geologists,* as a great difficulty on the ordinary theory, of granite having been injected whilst liquefied; but if we reflect on the probable state of the lower surface of a laminated mass, like clay-slate, after having been violently arched by a body of molten granite, we may conclude that it would be full of fissures parallel to the planes of cleavage; and that these would be filled with granite, so that wherever the fissures were close to each other, mere parting layers or wedges of the slate would depend into the granite. Should, therefore, the whole body of rock afterwards become worn down and denuded, the lower ends of these dependent masses or wedges of slate, would be left quite isolated in the granite; yet, they would retain their proper lines of cleavage, from having been united, whilst the granite was fluid, with a continuous covering of clay-slate.

Following, in company with Dr. A. Smith, the line of junction between the granite and the slate, as it stretched inland, in a S.E. direction, we came to a place, where the slate was converted into a fine-grained, perfectly characterized gneiss, composed of yellowish-brown granular feldspar, of abundant black brilliant mica, and of few and thin laminæ of quartz. From the abundance of the mica in this gneiss, compared with the small quantity and excessively minute scales, in which it exists in the glossy clay-slate, we must conclude, that it has been here formed by the metamorphic action,—a circumstance doubted, under nearly similar circumstances, by some authors. The laminæ of the clay-slate are straight; and it was interesting to observe, that as they assumed the character of gneiss, they became undulatory with some of the smaller flexures angular, like the laminæ of many true metamorphic schists.

Sandstone formation.—This formation makes the most

* See M. Keilhau's Theory on Granite, translated in the Edinburgh New Philosophical Journal, vol. xxiv. p. 402.

imposing feature in the geology of Southern Africa. The strata are in many parts horizontal, and attain a thickness of about 2000 feet. The sandstone varies in character; it contains little earthy matter, but is often stained with iron; some of the beds are very fine-grained and quite white; others are as compact and homogeneous as quartz rock. In some places I observed a breccia of quartz, with the fragments almost dissolved in a siliceous paste. Broad veins of quartz, often including large and perfect crystals, are very numerous; and it is evident in nearly all the strata, that silica has been deposited from solution in remarkable quantity. Many of the varieties of quartzite appeared quite like metamorphic rocks; but from the upper strata being as siliceous as the lower, and from the undisturbed junctions with the granite, which in many places can be examined, I can hardly believe that these sandstone-strata have been exposed to heat.* On the lines of junction between these two great formations, I found in several places the granite decayed to the depth of a few inches, and succeeded, either by a thin layer of ferruginous shale, or by four or five inches in thickness of the recemented crystals of the granite, on which the great pile of sandstone immediately rested.

Mr. Schomburgk has described† a great sandstone formation in northern Brazil, resting on granite, and resembling to a remarkable degree, in composition and in the external form of the land, this formation of the Cape of Good Hope. The sandstones of the great platforms of Eastern Australia, which also rest on granite, differ in containing more earthy and less siliceous matter. No fossil remains have been discovered in these three vast deposits. Finally, I may add that I did not see any boulders of far-transported rocks at

---

* The Rev. W. B. Clarke, however, states, to my surprise, (Geolog. Proceedings, vol. iii. p. 422,) that the sandstone in some parts is penetrated by granitic dikes: such dikes must belong to an epoch altogether subsequent to that, when the molten granite acted on the clay-slate.

† Geographical Journal, vol. x. p. 246.

the Cape of Good Hope, or on the eastern and western shores of Australia, or at Van Diemen's Land. In the northern island of New Zealand, I noticed some large blocks of greenstone, but whether their parent rock was far distant, I had no opportunity of determining.

APPENDIX.

DESCRIPTION OF FOSSIL SHELLS,
By G. B. SOWERBY, Esq., F.L.S.

SHELLS from a tertiary deposit, beneath a great basaltic stream, at St. Jago in the Cape de Verde Archipelago, referred to at p. 4 of this volume.

1. LITTORINA PLANAXIS. *G. Sowerby.*

Testâ subovatâ, crassâ, lœvigatâ, anfractibus quatuor, spiralitèr striatis ; aperturâ subovatâ ; labio columellari infimâque parte anfractûs ultimi planatis : long. 0·6, lat. 0·45, poll.

In stature and nearly in form this resembles a small periwinkle ; it differs, however, very materially in having the lower part of the last volution, and the columellar lip as it were cut off and flattened, as in the Purpuræ. Among the recent shells from the same locality, is one which greatly resembles this, and which may be identical, but which is a very young shell, and cannot therefore be strictly compared.

2. CERITHIUM ÆMULUM. *G. Sowerby.*

Testâ oblongo-turritâ, subventricosâ, apice subulato, anfractibus decem levitèr spiralitèr striatis, primis serie unicâ tuberculorum instructis, intermediis irregularitèr obsoletè tuberculiferis, ultimo longè majori absque tuberculis, sulcis duobus ferè basalibus instructo : labii externi margine interno intùs crenulato : long. 1·8, lat. 0·7, poll.

This species resembles so nearly one of the shells brought together by Lamarck, under the name of Cerithium Vertagus, that at first sight I thought it might be identical with it; it may be easily distinguished, however, by its being destitute of the fold in the centre of the columella so conspicuous in those shells. There is only one specimen, which has unfortunately lost the lower part of the outer lip, so that it is impossible to describe the form of the aperture.

3. VENUS SIMULANS. *G. Sowerby.*

Testâ rotundatâ, ventricosâ, lœviusculâ, crassâ; costis obtusis, latiusculis, concentricis, anticè posticeque tuberculatim solutis; areâ cardinali posticâ alterœ valvœ latiusculâ; impressione subumbonali posticâ circulari: long. 1·8, alt. 1·8, lat. 1·5, poll.

A shell which is intermediate in its characters, taking its place between the *Venus verrucosa* of the British Channel and the *V. rosalina* of *Rang* of the western coast of Africa, but sufficiently distinguished from both by its broad, obtuse, concentric ribs, which are divided into tubercles both before and behind. It is also of a more circular form than either of those species.

The following Shells, from the same bed, as far as they can be distinguished, are known to be recent species.

4. PURPURA FUCUS.

5. AMPHIDESMA AUSTRALE. *Sowerby.*

6. CONUS VENULATUS. *Lam.*

7. FISSURELLA COARCTATA. *King.*

8. PERNA,—two odd valves, but in such condition that it cannot be identified.

9. OSTREA CORNUCOPIÆ. *Lam.*

10. ARCA OVATA. *Lam.*

11. PATELLA NIGRITA. *Budgin.*

12. TURRITELLA BICINGULATA? *Lam.*

13. STROMBUS,—too much worn and mutilated to be identified.

14. HIPPONYX RADIATA. *Gray.*

15. NATICA UBER. *Valenciennes.*

16. PECTEN, which in form resembles *opercularis*, but which is distinguishable by several characters. There is only a single valve, wherefore I cannot consider myself warranted to describe it.

17. PUPA SUBDIAPHANA. *King.*

18. TROCHUS—indeterminable,

EXTINCT LAND-SHELLS FROM ST. HELENA.

The following six species were found associated together, at the bottom of a thick bed of mould; the two last species, namely, the *Cochlogena fossilis* and *Helix biplicata* were found, together with a species of *Succinea* now living on the Island, in a very modern calcareous sandstone. These Shells are referred to at p. 89 of this volume.

1. COCHLOGENA AURIS-VULPINA. *De Fer.*

This species is well described and figured in Martini and Chemnitz's eleventh volume. Chemnitz expresses doubts as to what genus it might properly be referred, and also a strong opinion unfavourable to the conclusion that it should be regarded as a landshell. His specimens were bought at a public auction in Hamburg, having been sent there by the late G. Humphrey, who appears to have been very well acquainted with their real locality, and who sold them for land-shells. Chemnitz, however, mentions one specimen in Spengler's collection, in a fresher condition than his own, and which was said to be from China. The representation which he has given is taken from this individual, and appears to me to have been only a cleaned specimen of the St. Helena shell. It is easy to suppose that a shell from St. Helena might have been either accidentally or interestedly, after passing through two or three hands, sold as a Chinese shell. I think it is not possible that a shell of this species could have been really found in China; and among the immense quantities of shells that come to this country from the Celestial Empire, I have never seen one. Chemnitz could not bring himself to establish a new genus for the reception of this remarkable shell, though he evidently could not collate it with any of the then known genera, and though he did not think it a land-shell, he has called it *Auris-vulpina*. Lamarck has placed it as the second species of his genus *Struthiolaria*, under the name of *crenulata*. To this genus it does not however bear any affinity; and there can be no doubt about the correctness of De Ferussac's views, who places it in the fourth division of his sub-genus *Cochlogena :* and Lamarck would have been correct according to his own principles, if he had placed it with his *Auriculæ.* A variety of this species occurs, which may be characterized as follows:—

Cochlogena auris-vulpina, *var.*

Testâ subpyramidali, aperturâ breviori, labio tenuiori: long. 1·68, aperturæ 0·76, lat. 0·87, poll.

Obs.—The proportions of this, differ from those of the usual variety, which are as follows:—Length 1·65, of the aperture 1·, width 0·96 inches. It is worthy of observation, that all the shells of this variety came from a different part of the island, from the foregoing specimens.

2. Cochlogena fossilis. *G. Sowerby.*

Testâ oblongâ, crassiusculâ, spirâ subacuminatâ, obtusâ, anfractibus senis, subventricosis, leviter striatis, suturâ profundè impressâ; aperturâ subovata; peritremate continuo, subincrassato; umbilico parvo: long. 0·8, lat. 0·37, poll.

This species is of the stature of *C. Guadaloupensis,* but may easily be distinguished by the form of the volutions and the deeply-marked suture. The specimens vary a little in their proportions. This species was not obtained by Mr. Darwin, but is from the collection of the Geological Society.

1. Cochlicopa subplicata. *G. Sowerby.*

Testâ oblongâ, subacuminato-pyramidali, apice obtuso, anfractibus novem lævibus, posticè subplicatis, suturâ crenulatâ; aperturâ ovatâ, posticè acutâ, labio externo tenui; columellâ obsoletè subtruncatâ; umbilico minimo: long. 0·93, lat. 0·28, poll.

This and the following are placed with De Ferussac's sub-genus Cochlicopa, because they are most nearly related to his *Cochlicopa folliculus.* As species they are, however, both perfectly distinct, being much larger, and not shining and smooth like *C. Folliculus,* which is found in the South of Europe and at Madeira. Some very young shells and an egg were found, which I conjecture to belong to this species.

2. Cochlicopa terebellum. *G. Sowerby.*

Testâ oblongâ, cylindraceo-pyramidali, apice obtusiusculo, anfractibus septenis, lævibus; suturâ posticè crenulatâ; aperturâ ovali, posticè acutâ, labio externo tenui, anticè declivi; colu-

mellâ obsoletè truncatâ, umbilico minimo; long. 0·77, *lat.* 0·25, *poll.*

This species differs from the last in being more cylindrical, and in being nearly free, when full grown, from the obtuse folds of the posterior volutions, as well as in the form of the aperture. The young shells of this species are longitudinally striated, and they have some very obsolete longitudinal folds.

1. HELIX BILAMELLATA. *G. Sowerby.*

Testâ orbiculato-depressâ, spirâ planâ, anfractibus senis, ultimo subtùs ventricoso, supernè angulari; umbilico parvo; aperturâ semilunari, supernè extùs angulatâ, labio externo tenui; interno plicis duabus spiralibus, posticâ majori: long. 0·15, *lat.* 0·33, *poll.*

The young shells of this species have very different proportions from those marked above, their axis being nearly as great as their width. The largest specimen is white, with irregular ferruginous rays. This is very different from any known recent species, although there are several to which it appears to have some analogy, such as *Helix epistylium* or *Cookiana,* and *H. gularis:* in both of these, however, the internal spiral plaits are placed within the outer wall of the shell, and not upon the inner lamina, as in *Helix bilamellata.* There is another recent species, which is somewhat analogous to this; it is as yet undescribed, and differs from this and from *Cookiana,* in the circumstance of its possessing four internal spiral plaits, two of which are placed within the outer, and two upon the inner wall of the shell; it was brought from Tahiti, in the *Beagle.*

2. HELIX POLYODON. *G. Sowerby.*

Testâ orbiculato-subdepressâ, anfractibus sex, rotundatis, striatis; aperturâ semilunari, labio interno plicis tribus spiralibus, posticis gradatim majoribus, externo intùs dentibus quinque instructo; umbilico mediocri; long, 0·07, *lat.* 0·15, *poll.*

This is somewhat related to *Helix contorta,* of De Ferussac, Moll. terr. et fluv. Tab. 51 A, f. 2; but differs from it in several particulars.

3. HELIX SPURCA. *G. Sowerby.*

Testâ suborbiculari, spirâ subconoideâ, obtusâ; anfractibus quatuor tumidis, substriatis; aperturâ magnâ, peritremate tenui; umbilico parvo, profundo; long. 0·1, *lat.* 0·13, *poll.*

Easily distinguished from *Helix polyodon*, by its wide, toothless aperture.

4. HELIX BIPLICATA. *G. Sowerby.*

Testâ orbiculato-depressâ, anfractibus quinque rotundatis, striatis; aperturâ semilunari, labio interno plicis duabus spiralibus, posticâ majori; umbilico magno; long. 0·04, *lat.* 0·1, *poll.*

This must be regarded as perfectly distinct from *Helix bilamellata*, on account of its form; its umbilicus is much larger, its spire is not flat, nor is the posterior edge of each volution angular. There are specimens, which must be referred to this species, found with the foregoing species, and with the *Coglogena fossilis*, which latter is associated with a living Succinea, in the modern calcareous sandstone.

PALÆOZOIC SHELLS FROM VAN DIEMEN'S LAND,

REFERRED TO AT P. 138 OF THIS VOLUME.

---

1. PRODUCTA RUGATA.

This is probably the same species with that named *Producta rugata* by Phillips (Geology of Yorkshire, part ii. plate vii. f. 16): it is, however, in too imperfect a condition to allow me to decide positively.

2. PRODUCTA BRACHYTHÆRUS. *G. Sowerby.*

Producta, testâ subtrapeziformi, compressâ, parte anticâ latiori, sub-bilobâ, posticâ augustiori, lineâ cardinali brevi.

The most remarkable characters of this species, are the shortness of the hinge-line, and the comparative width of the anterior part: its outside is ornamented with small, blunt tubercles, irregularly placed: it is in limestone, of the ordinary grey colour of mountain limestone. Another specimen, which I suppose to be an impression of the inside of the flat valve, is in stone, of a light, rusty-brown colour. There is a third specimen, which I believe to be the impression of the inside of the deeper valve, in a nearly similar stone, accompanied by other shells.

1. SPIRIFERA SUBRADIATA. *G. Sowerby.*

Spirifera, testâ lævissimâ, parte medianâ latâ, radiis lateralibus utriusque lateris paucis, inconspicuis.

The breadth of this shell is rather greater than its length. The rays of the lateral surfaces are very few and indistinct, and the medial lobe is uncommonly large and wide.

2. SPIRIFERA ROTUNDATA? *Phillips's Geology of Yorkshire,* pl. ix. f. 17.

Although this shell is not exactly like the figure above referred to, it would perhaps be impossible to find any good distinguishing character. Our specimen is much distorted; it is, moreover, an example of that sort of accidental variation that shows how little dependence ought, in some instances, to be placed upon particular characters; for the radiating ribs of one side of one valve, are much more numerous and closer than those on the other side of the same valve.

3. SPIRIFERA TRAPEZOIDALIS. *G. Sowerby.*

Spirifera, testâ subtetragonâ, medianâ parte profundâ, radiis nonnullis, subinconspicuis ; radiis lateralibus utriusque lateris septem ad octo distinctis: long. 1·5, lat. 2·, poll.

There are two specimens of this, in a dark, rusty, gray limestone, probably bituminous.

SPIRIFERA TRAPEZOIDALIS, var. ? *G. Sowerby.*

Spirifera, testâ radiis lateralibus tripartitim divisis, lineis incrementi antiquatis, cæteroquin omninò ad Spiriferam trapezoidalem simillimâ.

At first I hesitated to unite this to *Spirifera trapezoidalis,* but observing that at the commencement the radiating ribs were simple, and knowing that these are subject to variations, I have thought it best merely to distinguish this specimen as a variety.

There are several other, probably distinct, species of Spiriferæ, but as these are only casts, it is obviously impossible to give the external characters of the species. Since, however, they are very remarkable, I have thought it advisable to give a name, together with a short description of each.

4. Spirifera paucicostata. *G. Sowerby.*

Length equal to about two-thirds of its breadth; ribs few and variable.

5. Spirifera Vespertilio. *G. Sowerby.*

Breadth more than double its length, radiating ribs rather large, distinct, and not numerous; posterior inner surface covered with distinct punctulations in both valves.

6. Spirifera avicula. *G. Sowerby.*

The proportions of this species are very remarkable, inasmuch as it appears to have been nearly three times as wide as it is long; the radiating ribs are not very numerous, and the internal posterior surface of one valve alone (the large valve) has been punctulated. In its proportions, it resembles Phillips's *Spirifera convoluta,** but as our *Sp. avicula* is only a cast of the inside, its proportions are not so abnormal as those of *Sp. convoluta.*

A specimen which is very much pressed out of its natural shape, but which still appears to differ somewhat in its proportions, shows not only the cast of the inside, but also the impression of the outside; its radiating ribs are very irregular, and numerous, but it must be regarded as doubtful whether some of them be not principal and others only interstitial: their irregularity renders it impossible to decide.

* Geology of Yorkshire, Part 2. Plate IX. f. 7.

DESCRIPTION OF SIX SPECIES OF CORALS, FROM THE PALÆOZOIC FORMATION OF VAN DIEMEN'S LAND.

BY W. LONSDALE, Esq., F.G.S.

1. STENOPORA TASMANIENSIS, sp. n.*

Branched, branches cylindrical, variously inclined or contorted; tubes more or less divergent; mouths oval, divisional ridges strongly tuberculated; indications of successive narrowing in each tube, 1—2.

THIS coral, in its general mode of growth, resembles *Calamopora* (*Stenopora ?*) *tumida*, (Mr. Phillips, Geol. Yorkshire, Part II. Pl. 1, fig. 62), but in the form of the mouth and other structural details, the differences are very great. *Stenopora Tasmaniensis* attains considerable dimensions, one specimen being 4½ inches in length, and half an inch in diameter.

The branches have individually great uniformity in their circumference, but they differ with respect to each other in the same specimen; and there is no definite method of subdivision or direction of growth. The extremities are occasionally hollow; and one specimen, about 1½ inches in length, and half an inch in breadth, is crushed completely flat. The tubes, in the best exposed cases, have considerable length, springing almost solely from the axis of the branch, and diverging very gently till they nearly reach the circumference, where they bend outwards. In the body of the branch,

---

* Though the characters of this genus are unpublished, it has been thought advisable not to give them fully in this notice, a very few species only having been examined. The coral is essentially composed of simple tubes, variously aggregated and radiating outwards. The mouth is round or oblong, and surrounded by projecting walls, having along the crest a row of tubercles. The mouth originally oval is gradually narrowed (στενός) by a band projecting from the inner wall of the tube, and is finally closed.

M

the tubes are angular from lateral interference; but, on approaching
the outer surface, they become oval in consequence of the interspaces produced by the greater divergence. Their diameter is very
uniform throughout, with the exception of the narrowings near the
terminations of the full-grown tubes. The walls in the interior of
the branches were apparently very thin, but there is a relatively
considerable thickness of matter at the circumference. No traces
of transverse diaphragms have been noticed within the tubes.

Cases illustrative of the changes to maturity and final obliteration
in the oval termination of the tubes are rare, but the following have
been observed. Where the mouth becomes free and oval, the walls
are thin and sharp, and perpendicular within the tube. In some
cases they are in contact; but, in others, they are separated by
grooves of variable dimensions, in which very minute foramina or
pores may be detected. As the mouth approaches towards maturity, the grooves are more or less filled up, and the walls thicken, a
row of very minute tubercles being discoverable along the crest. At
this stage, the inner side of the tube ceases to be vertical, being
lined by a very narrow inclined band. The mature mouths are
separated by a bold ridge, generally simple, but not unfrequently
divided by a groove; the double as well as the single ridge being
surmounted by a row of prominent tubercles almost in contact
with each other. Only one example of the filling up of the mouths
has been observed, but it affords satisfactory evidence of a gradual
expansion of the inner band, before alluded to, and a final meeting
in the centre. In this extreme state, there is a general blending of
details, but the tubercles are for the most part distinct.

In this species, proofs of a narrowing of the mouth previously to
the formation of the perfect tube, and the final contraction, are not
very prominently exhibited in the long cylindrical straight branches;
but near the point where the tubes bent outwards, there is an annu-
lar indentation, which may be traced successively from cast to cast
in a lineal direction, parallel to the surface; and between the prominent narrowing and the perfect surface, the walls of the tubes
were slightly rugose. In another short branch, believed to belong
to this species, but in which the tubes diverged outwards very
rapidly, the narrowing is strongly marked, but not to an equal extent throughout the specimen.

The matrix, in which the fossil is imbedded, is a coarse calcareous shale, or a gray limestone; and in which occur also *Fenestella
internata*, &c.

2. Stenopora ovata, sp. n.

Branched, branches oval ; tubes relatively short, divergence great ; mouths round ; contractions or irregularities of growth numerous.

The characters of this species have been very imperfectly ascertained. The branches are not uniformly oval, even in apparently the same fragment. The tubes diverged rapidly along the line of the major axis, and had but a very limited vertical growth. Their casts exhibit a rapid succession of irregularities of development. The mouths, as far as they can be determined, were round or slightly oval, and the dividing, tuberculated ridges sharp; but in consequence of the outer surface not being exposed, their perfect characters, and the changes incidental upon growth, could not be ascertained.

The coral is imbedded in a dark gray limestone.

1. Fenestella ampla, sp. n.

Cup-shaped ; celluliferous surface internal ; branches dichotimous, broad, flat, thin ; meshes oval ; rows of cells numerous, rarely limited to two, alternate ; transverse connecting processes sometimes cellular ; inner layer of non-cellular surface very fibrous ; external layer very granular, non-fibrous ; gemmuliferous vesicle ? small.

Some of the casts of this coral have a general resemblance to *Fenestella polyporata*, as represented in Captain Portlock's Report on the Geology of Londonderry, Pl. XXII. A. fig. 1 a, 1 d ; but there is no agreement between the Van Diemen's Land fossil, and the structure of that species as given in Pl. XXII. figure 3. of the same work, or in Mr. Phillips' original figures, Geology of Yorkshire, Part 2, Pl. I. figures 19, 20. A general resemblance also exists between *Fenestella ampla* and a coral obtained by Mr. Murchison from the carboniferous limestone of Kossatchi Datchi, on the eastern flank of the Ural mountains, but there is again a marked difference in structural details.

Fenestella ampla attained considerable dimensions, fragments apparently of one specimen covering an area of $4\frac{1}{2}$ inches by 3 inches ; and it displays considerable massiveness of outline, the branches at the points where they dichotimose often exceeding the tenth of an inch in breadth.

In the general aspect of the coral a considerable uniformity prevails, but the branches vary in breadth, swelling out greatly near the bifurcations; nevertheless, there is no marked difference of character between the base and the upper part of the cup, even in the number of the rows of cells.

In the best state of the cellular surface, which has been noticed, the mouths of the cells are relatively large, round or oval, and are defined by a slightly raised margin; and an undulating, thread-like ridge winds between them, dividing the interspaces into lozenge-shaped areas. The rows of cells, immediately preceding the bifurcation, sometimes amount to ten, and after the separation generally exceed two. The mouths of the lateral rows project into the meshes; and the transverse connecting processes are sometimes cellular. The interspaces between the mouths, as well as the undulating ridges, are granular, or very minutely tuberculated. Internally, the cells exhibit the usual oblique arrangement, overlying each other, and terminating abruptly against the dorsal part of the branch. The perfect casts of the cellular surface give the reverse of the characters just noticed, but more generally the impressions display scarcely a trace of any other structure than longitudinal rows of circular mouths.

On the inner layer of the non-cellular surface, twenty well-marked parallel fibres, with intermediate narrow grooves or corresponding casts, may sometimes be detected, and the number is always considerable. The mode of preservation did not permit the true nature of the fibres to be discovered, but in consequence of what has been noticed in other species, it is inferred that they are tubular. Their range is considerable, but in the specimen, which exhibits their structure most fully, they are frequently cut off by circular foramina. Their perfect surface is minutely granular. The outer layer, or back of the branches, is composed of an uniform crust without any indications of fibres, but covered with numerous microscopic papillæ, and corresponding pores penetrating the substance of the layer.

The only indications of gemmuliferous vesicles, are small circular pits occasionally situated over the mouth, and agreeing in position with the vesicles, which in other cellular genera, have been considered as gemmuliferous. In the Russian specimen before alluded to, casts of similar pits are very uniformly distributed between the casts of the mouths.

The youngest state of the coral has not been noticed, nor have any marked changes incident upon age, except the gradual thickening of the non-cellular surface, by the coating over of the fibrous layer.

The matrix of the specimens is a dark grey splintery or an earthy limestone.

2. FENESTELLA INTERNATA, sp. n.

Cup-shaped; celluliferous surface internal; branches, dichotimous, compressed, breadth variable; meshes oblong, narrow; rows of cells 2—5, divided by longitudinal ridges; transverse connecting processes short without cells; non-cellular surface, inner layer, sharply fibrous, outer layer, minutely granular.

By the delicacy of its structure, this species is easily distinguishable from *Fin. ampla;* and in the rows of cells varying from two to five, as well as in their mode of development, there are further well-marked differences. It appears to have attained considerable dimensions, fragments having been noticed an inch and a-half in length and an inch in breadth.

The branches vary in width, swelling out gradually towards the bifurcations, but without any alteration in the form or size of the meshes; and as far as the state of the specimens will permit an opinion to be formed, no marked changes occurred during the development of the cup, except one about to be noticed. On the celluliferous surface of the branches, considerable, but uniform, alterations take place between the successive bifurcations. For a short distance above the point of separation, the branch is narrow and angular, and traversed along the centre by a ridge, and there is only one row of cellular mouths on each side. As the branch grew, the ridge widened, and ultimately became celluliferous, a row of mouths springing from its place (*internata*). The three ranges of cellular openings are, in this state of the branch, separated by two ridges, and these, as the development advanced, again widened and became cellular, the five rows being divided by four ridges. This appears to be the extreme stage of growth, another bifurcation taking place immediately after. In the earliest formed part of the cup only two or three rows of mouths prevail; and where the number is greater, a certain amount of irregularity in the linear arrangement is perceptible, resulting from the lateral expansion of the branch.

In the best preserved specimens, the mouths are relatively large, round or oval, and the margin is slightly raised. In the middle rows they are parallel or nearly parallel, and in the direction of the axis of the branch; but in the side rows they are often obliquely placed, inclining towards the meshes. In these nearly perfect specimens the dividing ridges are thread-like and slightly waved, but there is no trace of the lozenge-shaped compartments so distinctly exhibited in *Fenestella ampla*. The interspaces between the mouths are flat or slightly convex. In specimens less finely preserved, or deprived of the original surface, the mouths are not uniform in outline, and have no projecting margin. The dividing ridges are also relatively broader; and the whole surface, including the transverse connecting processes, is granular or minutely tuberculated.

The inner layer of the non-cellular surface is sharply fibrous, and the same structure may be more or less clearly detected in the transverse, connecting processes. The number of fibres on the branches do not apparently exceed twelve, and they are in general less numerous. Their range is considerable, additional ones being interpolated as the branch widens; and their surface is minutely tuberculated. No separate, circular foramina were noticed. The outer layer is uniformly granular, where completed, but every intermediate state from the sharply fibrous may be traced on the same specimen.

No distinct proofs of gemmuliferous vesicles have been observed, but in a specimen, which is believed to exhibit impressions of this species, there are occasionally to be detected, near the mouths, hemispherical casts, perfectly rounded on the surface, and evidently unconnected immediately with the interior of the cells, and which it is presumed may represent those vesicles. *Fenestella internata* appears to be an abundant fossil, one slab nearly eight inches long and six wide, being covered on both sides with fragments of it, and numerous smaller specimens occur in the collection. The matrix is chiefly a coarse gray calcareous shale, but it is sometimes a splintery limestone, or a hard ferruginous or light-coloured clay-stone.

3. Fenestella fossula, sp. n.

Cup-shaped, celluliferous surface internal; branches dichotimous slender; meshes oval; rows of cells, two; transverse processes non-cellular; inner layer of non-celluliferous surface minutely fibrous; external layer smooth or granular.

In general aspect and structural details, this species bears a great resemblance to *Fenestella flustracea* of the magnesian limestone of England, (*Retepora flustracea*, Geol. Trans., 2nd Series, Vol. III.

Pl. XII. f. 8), but it differs from it in the peculiar character exhibited in the cast of the celluliferous surface, the nature of which will be given in noticing that surface.

The principal specimen is a nearly perfect cup 1½ inches in height, and about two inches across the widest, compressed part. There are no marked variations of character, but occasionally, irregularities of growth, due, apparently, to accidents during progressive development.

The following details have been obtained from casts, no perfect surface having been noticed.—The branches had great uniformity of dimensions, swelling but very slightly at the distant points of bifurcation, and their thickness was apparently nearly equal to their breadth. The cast of the cellular surface is traversed along the centre by a sharp narrow trench (*fossula*), with nearly vertical sides, the distinguishing character between this species and *Fen. flustracea*. The cylindrical casts of the mouths, or the interior of the cells, are arranged in a single row on each side of the trench, and no increase of number is clearly perceptible at the bifurcations. Along the centre of the trench is a row of indentations or minute conical pits, a character noticeable in other species, particularly in *Fen. flustracea*. They are plainly not casts of cellular openings, but of relatively large papillæ. Traces of such projections have aso been noticed in several other instances.

The mouths of the cells, in the minute fragment which has been obtained exhibiting them, are large, round, slightly projecting, and not very distant, and in the same atom is an imperfect keel. The remains of the non-cellular surface exhibit no characters requiring notice, but indications of a striated and smooth layer have been observed.

The two specimens which afforded these structural details have a matrix of dark-coloured, hard limestone.

HEMITRYPA SEXANGULA, sp. n.

Net-work fine, hexagonal ; meshes round in double rows.

The coral to which the above inefficient characters are applied, is imbedded in the shaly surface of a dark, hard limestone. It is about an inch in breadth and half an inch in height, and consists of two layers of net-work,—one presenting quadrangular meshes, and the other hexagonal, with a round, inner area ; and over a considerable part of the specimen, the quadrangular network has been removed, whereby the connexion of the two structures is perfectly exposed.

This fossil is believed to agree completely in its essential generic characters with those of Hemitrypa (Pal. Foss. Cornwall, p. 27), but its state of preservation, and some facilities afforded by it for determining structural details, have led to an inference respecting its nature somewhat different from that given in the work just quoted.

The inner surface of *Hemitrypa oculata* (*loc. cit.*) is described as "marked with radiating ridges," having intervening "oval depressions, which penetrate only half through the substance of the coral, and no where reach the outer surface." The equivalent portion of the Van Diemen's Land specimen agrees perfectly with this statement, except in the form of the meshes or depressions; it is, however, not merely "like some Fenestellæ," but it possesses all the essential characters of that genus, and is believed to be a fragment of *Fen. fossula*. This inference is drawn from a minute portion mechanically detached, and which exhibited a row of large, round, projecting, cellular mouths. The external surface of *Hem. oculata* is described as "wholly covered with numerous round pores or cells"—"associated in double rows," and the corresponding portion of *Hem. sexangula* has been ascertained to consist also of a similar surface of double rows of round meshes or "pores," but with hexagonal boundaries; and they are shown, as exhibited by the specimen in its embedded state, to penetrate to the surface of the Fenestella or quadrangular net-work.

These details are conceived to be sufficient to establish a generic agreement between the Van Diemen's Land coral and *Hemitrypa oculata ;* and an examination of an Irish specimen of that genus has fully confirmed the structural details exhibited in the "inner surface" of the specimen to which, provisionally, the name of *Hemitrypa sexangula* is applied.

Of the true nature of the "external" net-work no opinion is ventured. It is formed almost entirely of dark gray, calcareous matter, filling apparently an originally cellular structure; but there are also a few small patches of the outer covering, consisting of an opaque white crust on the surface, which was originally in contact with the external net-work. That it was a parasite little doubt is entertained, and the interesting agreement between the space occupied by the double row of meshes, and that of the parallel branches of the Fenestella arises apparently from the latter having afforded suitable base lines for attachment. In the Van Diemen's Land specimen, the agreement is marked by an increased breadth in the net-work, and by a row of projecting points. There is also a re-markable agreement between the arrangement of the mouths of the Fenestella and the meshes of the "inner" net-work. Similar con-formities are admirably shown in Mr. Phillips's excellent figures (Pal. Foss. Pl. XIII. f. 38).

The solid portions of the structure being exceedingly fine, resem-bling the thread of the most delicate lace, attempts to discover satis-factorily internal characters proved unsuccessful, except in one place, where a true cellular arrangement was believed to be visible.* Of the nature of the investing crust, nothing also has been determined.

Though the name Hemitrypa may be objectionable, as applied to the corals under consideration, it has been thought right to retain the word, until the full characters of the genus shall have been ascertained.

FALMOUTH, *January*, 1844.

* A Codrington lens, half-an-inch in diameter, was invariably used in ex-amining the corals described in this notice.

INDEX.

ABEL, M., on calcareous casts at the Cape of Good Hope, 146
Abingdon Island, 104
Abrolhos Islands, incrustation on, 33
Aëriform explosions at Ascension, 39
Albatross, driven from St. Helena, 90
Albemarle Island, 103
Albite, at the Galapagos Archipelago, 104
Amygdaloidal cells, half filled, 27
Amygdaloids, calcareous origin of, 14
Ascension, 34
    arborescent incrustation on rocks of, 33
    absence of dikes, freedom from volcanic action, and state of lava-streams, 92
Ascidia, extinction of, 141
Atlantic ocean, new volcanic focus in, 92
Augite fused, 111
Australia, 130
Azores, 24, 125

Bahia in Brazil, dikes at, 123
Bailly, M., on the mountains of Mauritius, 29
Bald Head, 144
Banks' Cove, 103, 107
Barn, The, St. Helena, 76
Basalt, specific gravity of, 120
Basaltic coast-mountains, at Mauritius, 29
    at St. Helena, 80
    at St. Jago, 17
Beaumont, M. Elie de, on circular subsidences in lava, 102
    on dikes indicating elevation, 94
    on inclination of lava-streams, 93
    on laminated dikes, 70

Beudant, M., on bombs, 37, 39
    on jasper, 47
    on laminated trachyte, 67
    on obsidian of Hungary, 63
    on silex in trachyte, 14, 46
Bermuda, calcareous rocks of, 144, 147
Bole, 139
Bombs, volcanic, 36
Borz, St. Vincent, on bombs, 37
Boulders, absence in Australia and the Cape of Good Hope, 152
    of greenstone in New Zealand, 152
Brattle Island, 109
Brewster, Sir D., on a calcareo-animal substance, 54
    on decomposed glass, 132
Brown, Mr. R., on extinct plants from Van Diemen's Land, 140
    on sphærulitic bodies in silicified wood, 62
Buch, Von, on cavernous lava, 103
    on central volcanos, 127
    on crystals sinking in obsidian, 117
    on laminated lava, 66
    on obsidian streams, 64
    on olivine in basalt, 104
    on superficial calcareous beds in the Canary Islands, 88

Calcareous deposit at St. Jago, affected by heat, 3—7
    fibrous matter, entangled in streaks in scoriæ, 11
    freestone at Ascension, 49
    incrustations at Ascension, 50
    superficial beds at King George's Sound, 144
    sandstone, at St. Helena, 86
Cape of Good Hope, 148

London:
Printed by STEWART and MURRAY,
Old Bailey.

Printed in the United States
By Bookmasters